ARCTIC JUSTICE

Spaces and Practices of Justice series

Series editor: **Agatha Herman**, Cardiff University

The Spaces and Practices of Justice series focuses on the intersections between spaces and practices to provide innovative and important interventions on examples of real-world (in)justice. The series explores food justice, scholar-activism, social movements, gender, sexuality, race, childhood, labour, trade, domestic spaces, environmental relations and consumption to open out different approaches to questions of justice grounded within everyday experiences and spaces.

Forthcoming in the series:

The Practice of Collective Escape: Politics, Justice and Community in Urban Growing Projects
Helen Traill, June 2023

Environmental Justice and Peacebuilding: Integrating Nature in Policy and Practice
Rebecca Farnum, July 2023

Researching Justice: Engaging with Questions and Spaces of (In)Justice through Social Research
Agatha Herman and **Joshua Inwood**, November 2023

Find out more at
bristoluniversitypress.co.uk/spaces-and-practices-of-justice

ARCTIC JUSTICE

Environment, Society and Governance

Edited by
Corine Wood-Donnelly and Johanna Ohlsson

First published in Great Britain in 2023 by

Bristol University Press
University of Bristol
1–9 Old Park Hill
Bristol
BS2 8BB
UK
t: +44 (0)117 374 6645
e: bup-info@bristol.ac.uk

Details of international sales and distribution partners are available at bristoluniversitypress.co.uk

Editorial selection and editorial matter © 2023 Wood-Donnelly, © 2023 Ohlsson

The digital PDF and ePub versions of this title are available Open Access and distributed under the terms of the Creative Commons Attribution-NonCommercial-NoDerivatives 4.0 International licence (https://creativecommons.org/licenses/by-nc-nd/4.0/) which permits reproduction and distribution for non-commercial use without further permission provided the original work is attributed.

British Library Cataloguing in Publication Data
A catalogue record for this book is available from the British Library

ISBN 978-1-5292-2481-8 paperback
ISBN 978-1-5292-2482-5 ePub
ISBN 978-1-5292-2483-2 OA PDF

The right of Corine Wood-Donnelly and Johanna Ohlsson to be identified as editors of this work has been asserted by them in accordance with the Copyright, Designs and Patents Act 1988.

All rights reserved: no part of this publication may be reproduced, stored in a retrieval system, or transmitted in any form or by any means, electronic, mechanical, photocopying, recording, or otherwise without the prior permission of Bristol University Press.

Every reasonable effort has been made to obtain permission to reproduce copyrighted material. If, however, anyone knows of an oversight, please contact the publisher.

The statements and opinions contained within this publication are solely those of the editors and contributors and not of the University of Bristol or Bristol University Press. The University of Bristol and Bristol University Press disclaim responsibility for any injury to persons or property resulting from any material published in this publication.

Bristol University Press works to counter discrimination on grounds of gender, race, disability, age and sexuality.

Cover design: blu inc
Front cover image: Stocksy/Adam Sébire
Bristol University Press use environmentally responsible print partners.
Printed in Great Britain by CPI Group (UK) Ltd, Croydon, CR0 4YY

Contents

Series Preface · vii
List of Figures and Tables · viii
Notes on Contributors · ix
Preface · xiii

Introduction: Justice in the Arctic · 1
    Corine Wood-Donnelly and Johanna Ohlsson

1. Applying a Transnational Theory of Justice to the Arctic · 8
    Johanna Ohlsson
2. Responsibility of and for Structural (In)Justice in Arctic Governance · 21
    Corine Wood-Donnelly
3. A Relational View of Responsibility for Climate Change Effects on the Territories and Communities of the Arctic · 36
    Tracey Skillington
4. A JUST CSR Framework for the Arctic · 51
    Darren McCauley
5. Collective Capabilities and Stranded Assets: Clearing the Path for the Energy Transition in the Arctic · 66
    Roman Sidortsov and Anna Badyina
6. Mainstreaming Environmental Justice? Right to the Landscape in Northern Sweden · 81
    Tom Mels
7. Sacrifice Zones: A Conceptual Framework for Arctic Justice Studies? · 96
    Berit Skorstad
8. Planning for Whose Benefit? Procedural (In)Justice in Norwegian Arctic Industry Projects · 109
    Ragnhild Freng Dale and Halvor Dannevig
9. The Complex Relationship between Forest Sámi and the Finnish State · 124
    Tanja Joona and Juha Joona

10	FPIC and Geoengineering in the Future of Scandinavia *Aaron M. Cooper*	139
11	Overarching Issues of Justice in the Arctic: Reflections from the Case of South Greenland *Joan Nymand Larsen and Jón Haukur Ingimundarson*	154
12	Seeing Like an Arctic City: The Lived Politics of Just Transition at Norway's Oil and Gas Frontier *Anna Badyina and Oleg Golubchikov*	168

Conclusion: Making Connections between Justice and Studies of the Arctic 183
Johanna Ohlsson and Corine Wood-Donnelly

Index 190

Series Preface

Agatha Herman

Justice refers to a broad concern with fairness, equity, equality and respect. Just from the daily news, it is readily apparent how questions of justice or, in fact, the more obvious experiences of *injustice* shape our everyday lives. From global trade to our own personal consumption; living or dying through war and peace; access to education; relations in the workplace or home; how we experience life through a spectrum of identities; or the more- than- human entanglements that contextualize our environments, we need to conceptualize and analyze the intersections between spaces and practices of justice in order to formulate innovative and grounded interventions.

The Spaces and Practices of Justice book series aims to do so through cutting across scales to explore power, relations and society from the local through to international levels, recognizing that space is fundamental to understanding how (in)justice is relationally produced in, and through, different temporal and geographical contexts. It is also always practised, and a conceptual focus on these 'doings and sayings' (Shove, 2014) brings a sense of the everydayness of (in)justice but also allows for analysis of the broader contexts, logics and structures within which such experiences and relations are embedded (Jaeger-Erben and Offenberger, 2014; Herman, 2018).

References

Herman, A. (2018) *Practising Empowerment in Post-Apartheid South Africa: Wine, Ethics and Development*, London: Routledge.

Jaeger-Erben, M. and Offenberger, U. (2014) 'A practice theory approach to sustainable consumption', *GAIA*, 23(S1): 166–74.

Shove, E. (2014) 'Putting practice into policy: reconfiguring questions of consumption and climate change', *Contemporary Social Science*, 9(4): 415–29.

List of Figures and Tables

Figures

9.1	Lapp villages in northern Fennoscandia	129
9.2	The official Sámi Homeland area in Finland (three northernmost municipalities)	133

Tables

4.1	Arctic Environmental Responsibility Index, top and bottom 3 companies	53
5.1	Analytical framework for assessing the energy transition – the collective capabilities and energy justice perspective	74

Notes on Contributors

Anna Badyina is Research Fellow in Energy Justice and Transitions (SPRU – Science Policy Research Unit), University of Sussex Business School. Badyina is a human geographer with research and policy expertise in urban and sustainability studies. She is particularly interested in the relationships between spatial practices and social, economic and sustainability transitions.

Aaron M. Cooper is a PhD candidate, University of Stavanger Business School, Department of Accounting and Law. His research interests are International Environmental Law, primarily environmental protection in the Arctic, the risks of geoengineering to Indigenous peoples and developing multilevel governance frameworks to ensure minority peoples are adequately represented in these discussions.

Halvor Dannevig is Research Professor and Research Leader, Western Norway Research Institute. Professor Dannevig has worked on adaptation and vulnerability to climate change in Arctic communities. He studied how fishermen and farmers in the Lofoten region in Norway were affected by environmental and societal change and their capacity to adapt to these changes. He has also studied adaptation to climate change in Norwegian municipalities.

Ragnhild Freng Dale is Senior Researcher, Western Norway Research Institute. Freng Dale is a social anthropologist with a particular interest in energy, climate change and how local communities are affected by climate and social structures. She received her education in Britain and Norway, and obtained her doctorate at the University of Cambridge in 2019. Her doctoral thesis focuses on future narratives linked to oil and gas in Northern Norway and Sápmi, as well as how industrial developments are experienced and understood on a local level in Hammerfest, Northern Norway's first and so far the only petroleum village.

Oleg Golubchikov is Reader in Human Geography, Director of Postgraduate Research, at Cardiff University. Golubchikov has interests in urban geography and sustainability studies. His research and publications focus on the politics of urban and regional development, the policy and practice of sustainable urbanism, as well as on a spatial critique of energy/low-carbon transitions. He has long-standing research interests in the relationships between urban, social and political transformations under market transition. He also studies the governance of innovation with regard to smart cities, climate-neutral cities, and sustainable urban development. His research informs international policies.

Jón Haukur Ingimundarson is Senior Scientist at the Stefansson Arctic Institute and Associate Professor of Anthropology and Arctic studies at the University of Akureyri. His research foci include the political ecology of farming systems, cultural change, and adaptation in medieval to modern Iceland and Greenland; and human development, socio-economic transformations, and adaptive responses to changing environmental, climatic, and economic conditions in the circumpolar North.

Juha Joona is University Researcher at the Arctic Centre, University of Lapland. Joona is a legal scholar and his field of specialization is property law, environmental law and legal history. Regionally, his research interests are focused on land use rights in the Finnish and Swedish Lapland. His main research interest concerns local peoples' various land-use rights, especially in the area owned by the Finnish and Swedish states.

Tanja Joona is Docent in International law, an Arctic Five Chair in Just Green Transition, and a Senior Researcher at the Arctic Centre of the University of Lapland (2000–). She is currently the Finnish Institutional leader of the H2020 project JustNorth (2020–2023): Toward Just, Ethical and Sustainable Arctic Economies, Environments and Societies. Docent Joona's research interests focus mainly on the Arctic region, with comparative legal and political aspects of Indigenous Sámi society and especially issues dealing with traditional livelihoods, international human rights law and identity questions in the context of justice and equality.

Darren McCauley is Professor in the Management of International Social Challenges at Erasmus University Rotterdam (EUR). He is also academic director of the Erasmus Initiative and strategic research pillar of Global Social Challenges (GSC). His mixed methods research agenda focuses on investigating what a just transition to a low carbon future looks like from a global perspective. This has involved funded empircal research in the Arctic, sub-Saharan Africa, South East Asia, Europe and the US. He works closely with international organizations to co-develop research and ensure maximum impact from his work.

Tom Mels is Associate Professor in the Department of Human Geography ay Uppsala University. His research is concerned with power, justice and the politics of nature and landscape, with a historical, materialist and humanist penchant. Empirical interests include capitalist modernity and nature conservation in Sweden; the production of land reclamation spaces on Gotland; and wind power landscapes.

Joan Nymand Larsen is Professor, University of Akureyri, Senior Scientist and Research Director, Stefansson Arctic Institute, and Adjunct Professor, University of Greenland. Her main research areas centres on economics, climate change impacts and adaptation, and Arctic human development; social indicators: analysis of Arctic human development, living conditions and quality-of-life in the North; and Arctic youth.

Johanna Ohlsson is Researcher in Ethics at the Institute for Russian and Eurasian Studies, Faculty of Social Sciences at Uppsala University, and an Assistant Professor at University College Stockholm, Department for Human Rights and Democracy. Her research interests cover human rights, sustainability, models of justice and political processes.

Roman Sidortsov is Associate Professor in Energy Policy at Michigan Technological University, and Senior Research Fellow in Energy Justice (SPRU – Science Policy Research Unit), University of Sussex Business School. Sidortsov has a diverse international background as an educator, researcher, consultant and practising attorney. He has developed and taught law and policy courses ranging from Renewable Energy and Alternative Fuels to Administrative Law at Vermont Law School, Irkutsk State Academy of Law and Economics in Russia, and Marlboro College Graduate School's MBA in Managing for Sustainability program.

Tracey Skillington is Director of the BA (Sociology) and Chair of the Undergraduate Committee in the Department of Sociology and Criminology at University College Cork. She is currently researching issues of justice that arise in relation to global climate change, cosmopolitanism and transnational democracy. Skillington's research interests include critical theory, climate justice, human rights, global justice, intergenerational inequalities, border practices, sociology of the body, European identity, cosmopolitanism, collective memory, trauma and models of democracy.

Berit Skorstad is Professor at the Faculty of Social Science, Nord University. Professor Skorstad is a sociologist and has research interests in ethics, environment, sustainability, Arctic communities and social science theory. She has been leading a project on Urban Mining and has research expertise on the connections between mining activities, knowledge politics and the valuation of landscapes in the Arctic.

Corine Wood-Donnelly is Associate Professor of Social Sciences at the Faculty of Social Sciences at Nord University. She is also a researcher at Uppsala University where she is the Scientific Coordinator for the EU-funded project JUSTNORTH (GA 869327). Wood-Donnelly is an interdisciplinary researcher in International Relations and political geography, and specializes in governance and policy of the Arctic region.

Preface

There is a lengthening history underpinning the development of the work and network of scholars presented in this volume. The first seed for studies of justice and the Arctic was inadvertently planted by Professor Sverker Sörlin at the June 2017 Ninth International Congress on Arctic Social Science (ICASS IX) in Umeå, Sweden, while speaking in a panel discussion on *Past Theories/Future Theories? A Round Table on 'Theory' and Arctic Social Science and Humanities*. In this discussion, the question was posed 'Is there a theory of justice on the Arctic?' As it turns out, there was not one then and we have not arrived at one in this volume either. However, there is now a small body of work on justice and the Arctic.

With the Arctic situated at the crossroads of colonial legacies, geopolitics, resource extraction, sustainable development, evolving governance and, in fact, many other global concerns such as climate change, pollution and ocean acidification – just to name a few – it behoves that the issues of justice and injustice be brought to the fore. Currently, many states are facing a reckoning with past practices of exploitation and marginalization that frame the context for the injustices of today. Yet despite these contestations for recognition, the need to adjust the distribution of harms and benefits and to improve decision-making procedures that affect communities, resources, economies and environments of the Arctic, there is little indication that the necessary change is on the horizon.

It is hoped that this work will accelerate scholarship and have an impact on decision making in the Arctic, and we are hopeful that change is on the way – not least due to growing interest in the united themes of the Arctic and justice. For example, all of this has been made possible by a number of funding organizations that have made generous provisions in supporting this conversation and research on Arctic justice. This includes funding for several workshops from the Uppsala Forum on Democracy, Peace and Justice and the International Arctic Science Committee, a network grant from the Centre for Integrated Research on Culture and Society at Uppsala University (funding both workshops and the open-access conditions of this volume), and finally, several chapters were made possible through research financing from the EU Commission's Horizon 2020 programme JUSTNORTH project (Grant Agreement 869327). The editors have also been generously supported by Michael Shirley in the copy-editing of this volume.

The question of 'Is there a theory of justice on the Arctic?' remains for various reasons, unanswered. However, the contributions in this volume certainly begin to signal what might need to be considered if such a theory is to be conceived.

The lengthy work of framing and understanding justice (and injustice) in and for the Arctic has begun. It is up to you to carry it forward into the future.

Corine Wood-Donnelly
Bodø, Norway

Introduction: Justice in the Arctic

Corine Wood-Donnelly and Johanna Ohlsson

Policy makers and scholars often see the Arctic as an attractive laboratory for international cooperation, especially concerning sustainable development and environmental protection, yet it is hardly considered a site for fostering or testing the principles or perspectives of justice. Across scales of geography and hierarchies of power, the conditions of the Arctic has provided repeated opportunities to generate new ideas for cooperation and equitable systemic structures that seek to redress injustice in the future development of the region, and, beyond development, a flourishing existence. Yet the chapters in this volume are testaments to the opportunities missed in establishing the just conditions found within theories of justice. We are aware that, despite the potentially good intentions held by stakeholders with interests in the region, there is an inherent disjunction between the governance of the Arctic and its economic development. Caught between this disjunction are the people and the environment that it affects – an environment that is increasingly connected to, and important for, the entire global system.

Despite its oftentimes intuitive conviction and common-sense use, there is not one definition for the notion of justice. Justice has been discussed throughout centuries, across civilizations and the globe – not only in legal, moral and political philosophy but also in the various disciplines of social sciences, and, more recently, in environmental sciences. This provides us with a broad and nuanced understanding of justice, but a commonly agreed upon definition remains elusive. Some emphasize justice as a normative concept, while others see it as a relational process, yet others view it as fundamental to societal structures. Some focus on issues of distribution, others on representation, participation and recognition, while others take up the concept of core values in framing moral positions. These concepts each make up the core of our understanding of justice. The understanding of justice will, therefore, have subjective meanings that will also depend on where one begins their analysis. That justice can be understood and operationalized differently becomes inevitably clear when exploring issues of justice and injustice in the Arctic. What justice is, could be and should be in the context of the Arctic is the theme of this book, and it starts an important conversation bridging research from various traditions. As an introductory collection of justice scholarship on the Arctic region, the purpose of this volume is rather to show the diversity of the notion of justice in the Arctic than to develop a grand theory of justice for the Arctic.

This volume marks the beginning of an inquiry into the issues and challenges of the Arctic through insights drawn from theories and perspectives of justice. In studies of the Arctic, the notion of justice is largely absent in normative, empirical and applied research. This void has existed even though scholars have long referred to injustices and problems in the region from the results of colonial legacies to the incorporation of the region into the international political economy, and even more recently, as the locus of concern for bearing an uneven impact of global climate change. Across society, environments and governance arrangements in the Arctic, there is momentum rising for recognition, reconciliation and transformation to ensure that the future of the region is not the same as its history. The purpose of this work is to introduce the ideas and theories of justice into the domain of Arctic research and to plant a seed for scholarship that makes investigations into how and why the conditions that foster injustice prevail in the Arctic.

Arctic Justice: Environment, Society and Governance offers a fundamental introduction to the study of different aspects of justice in the Arctic region. The chapters all contribute to the understanding of justice in the Arctic and, to varying degrees, consider three overarching themes: (1) global or broader circumpolar contexts to local challenges facing communities, (2) responsibility for justice in governance on various administrative levels, and finally, (3) failures of justice in distribution, procedures and recognition within the environment, society and territory. This work provides foundational insights for justice research on the Arctic and marks the starting point for such future research. Inspired by key thinkers in various traditions of justice theories, these chapters highlight the practical consequences of Arctic imperialism, resource exploitation and unequal power hierarchies in its governance practices.

Each individual critique draws the reader's attention to the familiar stories of the Arctic: global warming, resource extraction, economic and governance development, but does so through the lens of the concept of justice. The scholars included in this volume range from experts in the concept and conceptions of justice to those with decades of experience in Arctic research. Sincere, scholarly and informative, the essays in this volume offer important insights and provide a fascinating overview of perspectives and possibilities for bridging the gap between the Arctic, as it is today, and a future Arctic with just conditions. The cases and foci included here are intended to draw attention to issues of justice at the heart of the Arctic region and begin with discussions of the broad ranging injustice that spans the entirety of the region, then narrows to more specific investigations on issues within a nation or a particular geographic place.

When it comes to issues of justice (and injustice) in the Arctic, many scholars have emphasized the climate change related challenges emerging in the Arctic. Additionally, aspects of socio-economic inequalities (see for instance Chapters 11 and 12), land dispossession (Chapter 9), resource grabbing (Chapter 8) and the colonial past (Chapters 9 and 11) still influence relations, structures and policies, and pose central questions for understanding issues of justice and injustice in the Arctic. Perspectives of justice theory could hence be seen as central to understanding the dynamics of justice and injustice in the Arctic region, with historical legacies central to comprehensively understanding contemporary circumstances. This contributes to the argument that justice theories – by addressing injustices – offer a crucial lens for increasing social, economic and environmental sustainability in the region. However, the use of the lens of justice theory has until now been lacking in Arctic research. This volume speaks directly towards this void.

Red threads of justice

The overarching concern of this book asks what we can learn about the Arctic when we apply the theories and ideas of justice to the region. This book investigates what the Arctic looks like through the lens of justice. The contributions include approaches from different disciplinary perspectives present within the group of scholars contributing chapters to this volume, drawing on justice as a conceptual tool in framing the ontology of the spatial and temporal relationships inherent in studies of the Arctic. The scope of this volume focuses on research that considers the environment, society and governance in the Arctic through the themes of responsibility and failures of justice through circumstances and conditions of injustice. Contributing to the existing discourse of normative and applied theories of justice, the volume seeks to conceptualize the role of justice in Arctic research, as well as find ways to promote and assist the transition from current modes of economic development and consumption towards a just and sustainable future. In claiming this broad academic scope on justice, the volume aims to speak to aspects of responsibility, both globally, transnationally and locally, as well as circumstances and conditions of injustice both normatively, conceptually and applied.

There are four central key features in this volume, which tie all of the chapters together. These are (1) the *Arctic*, (2) *normative aspects in conjunction with empirical problems of injustice*, (3) the centrality of the concept of *rights*, and (4) a *multidisciplinary* approach. Firstly, the initial key feature is the Arctic region. By bringing the Arctic and justice together, the volume contributes new insights of relevance for several disciplines, such as various environmental social sciences, but also policy makers. Secondly, this volume contributes to the existing literature by bringing in normative aspects of justice in conjunction with issues of ethics, which arise in empirically based problems of injustice. By exploring what a just and ethically defensible future for the Arctic could and should look like, the volume benefits the reader in that it combines normative and empirical research, often by looking at conditions created by the past. Thirdly, in much of the authors' work throughout the chapters, the notion of rights holds a central position. This volume offers readers a nuanced account of what questions different types of rights give rise to. This volume's multidisciplinary approach is what makes this research possible and serves as the fourth key feature. The authors utilize sociology, geography, law, International Relations, political science, anthropology and ethics to create their arguments. This offers the reader a comprehensive volume which bridges theory and empirical work while offering a novel way of addressing and understanding issues in the Arctic.

Structure of the volume

The volume is organized so that the contributions and the questions posed by the scholars start from an international and conceptual perspective, and then continue toward a regional, national and community oriented one based on applied and empirical scrutiny in the various case studies. The first four chapters have a broader conceptual outlook centred on the overarching theoretical and empirical issues of justice and injustice in the Arctic, primarily focusing on aspects of responsibility. The chapters address what responsibility could (or should) look like in the Arctic in terms of environmental, societal or governance approaches. Chapters 1, 2, 3 and 4 focus on international and transnational aspects of justice in the Arctic, and level critique towards previous as well as current structures, based on theories of justice.

Chapter 1 applies Rainer Forst's theory of transnational justice to the region's norms, sovereignty and development. It specifically considers the issue of Arctic exceptionalism. By assessing Forst's normative principles and relating them to the structure and set up of the Arctic Council, the chapter contributes to the discussion on agency and governance in the Arctic. Within the framework of the International Relations theory of Social Constructivism, Chapter 2 explores the function of the Arctic Council through the lens of Iris Marion Young's conceptions of structural injustice, five faces of oppression, and designations of responsibility for (in)justice. By showing where existing rules of governance result in oppression, this chapter contributes to a discussion of the role and potential of the Arctic Council and governance in the region more broadly.

Chapter 3 contributes to the understanding of climate justice in general, and the particular challenges facing people, nature and landscapes in the

Arctic. By taking stock of what might be perceived as relevant normative standards for taking responsibility for the effects of rising global temperatures on the territories and communities of the Arctic, the chapter discusses the conditions for responsibility and accountability. Further, the chapter proposes a relational 'civic connections approach' model of responsibility that emphasizes the interconnectedness of peoples, regions, climate actions and outcomes.

In looking at the transnational scope of Arctic development, Chapter 4 contributes a critical discussion on corporate social responsibility (CSR), suggesting that taking responsibility principally is not enough and that companies also need to consider a variety of justice concerns. The chapter argues that the processual focus of responsibility tends to leave out important aspects of the outcomes and their consequences, and thus why it is important to adhere to both processes and outcomes, combining various justice theories. Putting forward the JUST (Justice, Universal, Space, Time) framework, the chapter contributes constructively to debates on the role and responsibilities of energy companies in the Arctic.

Contributing important discussions on circumstances and conditions of injustice in the Arctic, and the assessability of the same, Chapters 5, 6 and 7, take more of an explicit applied justice approach into consideration, centring on the fields of energy and environmental justice. Chapter 5 is positioned in the emerging tradition of energy justice scholarship and speaks to some debates in the literature on Just Transitions. Here, the authors make novel and important contributions to the utility of the Capabilities Approach in conceptualizing and assessing the impact of oil and gas activities on the energy transition in the Arctic.

Chapter 6 takes up the adoption and mainstreaming of the concept of environmental justice into various legal and policy instruments, such as the European Landscape Convention (2000), the global Sustainable Development Goals (2015) and, more recently, the European Green Deal (2019). Voicing a crucial concern about these practices only leading to signposting sustainability rather than creating actual social and political change, the chapter argues that such mainstreaming ultimately depends on a narrow and idealist theorization not simply of justice, but of nature and history in the production of landscapes. Chapter 7 also relates to aspects of environmental justice. It addresses and assesses the concept of Sacrifice Zones. This makes an important contribution by expanding the discourse on environmental justice as well as Sacrifice Zones beyond the American context. By assessing the viability of the concept and linking it to aspects of justice and injustice in the Arctic, the chapter contributes an innovative and unifying framework that is helpful for analyses that examine the linkages between the environmental and human challenges in industrial extractive projects in the Arctic.

Several chapters in the volume, particularly Chapters 8, 9 and 10, analyse various issues and challenges facing Indigenous peoples – both generally and with specific attention to the Indigenous populations in Finland and Norway. For instance, Chapter 8 explores how industry projects affect Sámi peoples in the Norwegian Arctic. The regulatory processes for initiating industrial projects in the Norwegian Arctic are extensive. Still, as the authors convincingly argue, rights holders (typically Indigenous peoples) often do not know when or where in the processes their voices will be heard – or desires actioned. Procedures can be developed 'by the book' but still leave significant room for interpretation which then creates ambiguities for representation, participation and recognition. In addition, stakeholders' and rightsholders' legitimate claims for having their rights respected are often reduced to 'interests', indicating a loss of trust in the state, but also a failure of the state in fulfilling the obligations inherent in the social contract.

Chapter 9 also contributes research based on Arctic Indigenous communities, and explores the historical legal context and colonial history that shapes the current debates and challenges for Forest Sámi in Northern Finland. Pinpointing how the legacy from the past still affects the opportunities of Indigenous peoples to secure their political autonomy, territory and cultural continuity, the chapter makes a crucial contribution towards our understanding of the Forest Sámi as the existing literature is scarce. Directly related to issues of Indigenous representation and the rights of Indigenous peoples, Chapter 10 critically discusses the role of free, prior and informed consent (FPIC) in geoengineering in the Arctic. By focusing on the emerging issues of solar radiation management and ice-geoengineering through procedural aspects of justice, the chapter utilizes ideas emerging from intergenerational justice.

Chapters 11 and 12 centre on aspects of justice and injustice of economic initiatives and developments in various parts of the Arctic. Economic extractive projects are now operating more widely in the Arctic, and Chapter 11 contributes a discussion on the circumstances, structures and colonial legacies of industrial projects in South Greenland. Being closely connected to Denmark, local solutions in Greenlandic society indicate the interconnectedness of many issues such as aspects of representation, participation and recognition. The chapter connects the local communities' legitimate demands for sustainable livelihoods with external interests and global trends.

Related to economic consequences and situated in a discussion on Just Transitions, Chapter 12 discusses the challenges met by local communities when extractive oil and gas industries establish themselves in remote areas of the Arctic, centring on Hammerfest in Northern Norway. This chapter indicates that the oil and gas industry has a central role to play in developing small towns and in creating opportunities for thriving societies, but that these industries must pursue these developments in a respectful and just way. It also conveys the complex picture of some societies thriving and others shrinking due to economic development trajectories.

Positions of justice

Taken together, the various chapters contribute to the volume being descriptive, explorative and normative in character. It verges on prescriptive recommendations for introducing change. They contribute to the different aspects of justice theory, as well as its applicability to the Arctic region. The chapters discuss injustices faced in the Arctic, and to various degrees relate these to theories or concepts of justice. Some of the chapters offer deductive discussions most often based on empirical and practical problems, yet with a primary theoretical focus, whilst others are more inductive, showing how the injustices empirically and contextually shape the societies, environments and territories of the Arctic. Most chapters in the volume contribute empirically grounded accounts, often with normative implications of high policy relevance for the public as well as business sectors.

Justice and injustice in the Arctic are multi-layered and multi-faceted. Each of the situations under study in this volume could have been examined from other perspectives of justice that may have come to different conclusions. There is also the question of subjectivity and positionality that emerges within this scholarship. Some of these assessments and analyses are conducted from an outsider perspective, yet often in co-creation with Arctic Indigenous communities, and most of this work is by scholars making no claims to Indigenous positions themselves. We acknowledge that this is a central and sensitive issue related to questions of Indigenous agency and recognition – and absolutely justice. The scholars in this volume are acutely aware about the debate on who legitimately could and should be researching Indigenous issues. Yet, speaking out on injustice and pointing towards the possibility for justice is the responsibility of scholars everywhere, regardless of cultural identities or global location.

1

Applying a Transnational Theory of Justice to the Arctic

Johanna Ohlsson

Introduction

This chapter addresses what theories of justice may help further our understanding of injustices in the Arctic. The purpose is to critically discuss the baseline for a Forstian transnational theory of justice and its applicability to the Arctic, primarily the Arctic Council. This will take into account the regional, political, Indigenous and environmental aspects of this specific region. The account suggested here draws primarily on Critical Theory, and the suggested approach proposes that there are normative criteria required for a comprehensive theory of Arctic justice and that these are of general, rather than regional, character. Hence, the chapter tests to what extent a transnational theory of Arctic justice is reasonable and posits scepticism towards a theory based on Arctic exceptionalism. It instead argues for a critical theory grounded in universal principles that embraces breadth and contextual sensitivity. The chapter contributes a discussion on the normative principles necessary for developing justice theorizing applicable to the Arctic region, as well as a discussion of the implications of assessing justice in a transnational context.

The chapter contributes a critical assessment of a few aspects of justice, which are relevant for approaching the concept of justice in the context of the Arctic. Such an assessment could be done in numerous ways, but one promising and underexplored avenue concerning the Arctic is an assessment of a Forstian notion of justice as the right to justification in Rainer Forst's transnational account of justice. The existing literature contributing to understandings of justice and injustices in the Arctic seems to primarily do so from a development or energy perspective, often adopting versions of

Amartya Sen's and Martha Nussbaum's capability approach (for example, Rauschmayer et al, 2015; Willand and Horne, 2018; Kortetmäki, 2018; Sidortsov and Badyina, 2023, this volume). As the first section of this volume aims to cover the characteristics of the overarching issues of justice and injustice in the Arctic, it seems an apt intellectual exercise to extend the exploration of existing justice scholarship and its application to this region. This justifies taking this position in Critical Theory as its starting point.

After this introduction, and before turning to a presentation, discussion and assessment of the Forstian theory of transnational justice, this chapter will highlight some of the aspects of justice and injustice that appear central to the Arctic. This is done through the lens of transnational relations, and with social justice in mind. The chapter then ends with a few concluding remarks and a set of study questions.

Issues of justice – and injustice – in the Arctic

The Arctic is a topical region for several reasons, not the least when it comes to issues related to justice. Issues of justice are of crucial importance in this region, yet on different scales covering the interaction between states, within states, as well as within and between groups and local communities. However, issues of justice in the Arctic have, up until now, received relatively scarce scholarly and policy interest. A few aspects which seem to be of primary importance are related to the distribution of power, influence and issues of recognition.

Concerning interstate relationships, some scholars have predicted that the region will be the new hotspot for international conflicts and great power struggles. These predictions have been made as tensions between Russia and the United States, which are separated by less than three miles in the Arctic, continue to grow (Crawford, 2021). Potential incompatibilities also concern access to, and distribution of, natural resources such as oil and gas (Keil, 2014), as well as conflicting interests between different economic sectors, such as mining and tourism (Similä and Jokinen, 2018). Other examples include conflicts between extractive industry and primary livelihood (fisheries, herding, hunting), and tensions around resources in the North Atlantic (Greenland, Svalbard and so on). Yet, at the same time, scholars have also argued that the Arctic is a region with high geopolitical stability (Heininen, 2018), where a changing Arctic would not be a realistic trigger for Great Power conflicts (Tunsjø, 2020). Additionally, the Arctic has been one of few areas in the world where the most powerful states have had continuous dialogue, and where the US and Russia have had a strong track record of cooperation (Pincus and the Foreign Policy Research Institute, 2021). For instance, in January 2021, the US and Russian coastguards carried out joint maritime border controls in the Bering Sea, indicating, supporting and demonstrating mutual agreements

(SeaPower, 2021). This changed after Russia's invasion of Ukraine, yet still paints a complicated picture of complex relations in the international and transnational context of the Arctic – a picture that relates to different aspects and types of justice and injustices in different ways. Notably, a perceived injustice is a common ground for increased tensions between actors. This tension sometimes escalates to conflict.

Issues of justice in the Arctic could mean a lot of different things and needs to be addressed on different levels. For instance, justice concerns will be understood and perceived differently depending on whether an international institutional approach or a local or Indigenous approach is at the centre of analysis. For instance, potential conflicts between interests and rights are identifiable in relations between industry and livelihood. This becomes particularly apparent if we differentiate between scales in, for instance, social, legal or political justice. This chapter will detail how justice in the Arctic could be understood on an international, primarily transnational, level. It discusses issues related to social and political justice through concepts such as recognition and representation. Here, principles such as generality, reciprocity as well as sovereignty and the 'all affected' principle play a central role. The reasoning here is positioned within the Critical Theory tradition, following Frankfurt School scholar Rainer Forst, who understands justice as a concept of non-domination and emancipation, deeply connected to the right to justification (Forst, 2014, pp 2–9, Forst 2001, p 120).

Some of the issues of injustices in the Arctic brought forward by Coggins et al (2021) are centred on persons, while others are directed towards groups, peoples or states. These offer different units of analysis and sometimes overlap. Pronounced inequalities are challenges faced both between persons, peoples and states, as well as within groups of peoples and states. The issue of land dispossession is often disputed between groups, but perhaps primarily within states. Further, issues raised by colonialism and colonial legacies could be of interstate as well as intrastate character while simultaneously affecting peoples and persons. This indicates that there are several structural levels at play when discussing issues of justice and injustice. These could be interpersonal relations between groups as well as relations between states, yet they always have a relational grounding. Hence, one could argue that local, national and regional aspects cut across several issues of justice in the Arctic and that the scope and agency of political actors vary. The focus in this chapter is primarily on a transnational level, acknowledging states as central actors, yet it also acknowledges persons and peoples as subjects of justice. On this transnational level, one platform for collaboration seems to be of particular importance for issues of Arctic governance, but also for addressing issues of justice and injustice: the Arctic Council. Hence, this chapter focuses on the Arctic Council as a platform for justice in the Arctic and uses this as an example for assessing the normative principles in Forst's theory of transnational justice.

Forst's theory of justice in transnational settings

The German political theorist and philosopher Rainer Forst offers a theory of justification where he argues that the two formal criteria of reciprocity and generality should be guiding the justness of any action, relationship or structure (Forst, 2014, 2016, 2017, 2020). His theory of justification is also his theory of justice. This theory seems to be primarily developed with the national context in mind, focusing on constitutional nation-states; however, Forst's theory is extendable to a transnational context (Forst, 2001), which is why is relevant for a theory of justice in the Arctic. For Forst, justice is a notion of high political relevance. He argues that: 'political and social justice is an autonomous collective process of producing social and political conditions that are not only susceptible of justification in reciprocal and general terms, but can themselves be established via justification and aim to realize a basic structure of justification' (Forst, 2017, p 9).

Forst discusses normative orders in relation to societies, which appears to be an analogy for the nation-state, given that he often takes a constitutional perspective. Following the idea of the nation-state as a normative order, the members of that order are to be seen as normative authorities. However, he also explicitly addresses transnational (in)justice, and argues that this has a normative basis in a democratic conception of justice, which is yet realistic (speaking to the tradition of realism in International Relations scholarship). He critiques parochial or positivistic conceptions of justice, as these are insufficient when questioning empirical and normative premises as they tend to focus on state-centric approaches and therefore miss the forms of injustice beyond the state (Forst, 2020). This clearly shows the importance of taking non-state centric accounts, such as transnational injustice, seriously. Forst, when discussing justice, talks about persons and the crucial aspect of individuals being free, equal (free from domination) and having the right to justification. This focus on persons is, as he seems to argue, extendable to groups of people. One interpretation of Forst is that he is a cosmopolitan constitutionalist, as one's right to justification is important also in a transnationalist setting. The right to justification sets two normative criteria: reciprocity and generality. These are the baselines for how to understand and assess justification. In other words, an action is justified if this is done in a reciprocal and general way. He states that: '*Reciprocity* means that no one may make demands that he or she denies to others and no one may impose his or her non-generalizable views, interests or values on others. *Generality* means that all those for whom norms claim to be valid have to be equally involved' (Forst, 2016, p 14).

However, Forst also develops these arguments by extending them, with transnational aspects being taken into consideration. Here, he argues that the same principles should also be governing transnational relations and that all affected people should be included in the processes of decision making. This is an inherent democratic argument, stressing the need for the peoples' potential to influence political decision making. Yet, it seems reasonable to include some aspects of a proximity principle in this equation. Even if the 'all affected' principle should govern the processes, it seems reasonable that the people living in the region should have a larger say in decision-making processes. But what is the basis for that argument? One foundation could be that it is the people in the region's everyday life that are most affected. But, on the other hand, how do we know that it is not also the everyday life of people living in small island states threatened by flooding who will be affected by measures in the Arctic? The effects simply might not be immediate. This illustrates a political and institutional challenge of the 'all affected' principle.

Based on the reasoning of Forst, a transnationalist structure is developed based on the moral aspect of all persons, as well as the political aspects of the cooperation between peoples. Here, Forst builds his reasoning on the Rawlsian notion of peoples, yet argues that the moral status of all persons should be seen as the same – regardless of whether a national or international perspective is applied. What is different is the political solution. This makes Forst's theory cosmopolitan. A common objection to the political ability to implement this universal human value (or principle of human dignity) is due to the fact that nation-states are still the main players in the international system, and that it is up to a nation-state's will and ability to act according to this universal moral status. However, the system of states tends always to prioritize their own citizens rather than the interests of people in general. Forst argues that a theory of justice must be 'realistic' in the right way, in that it has to be receptive 'when it comes to assessing the current world order as one of multiple forms of domination' (Forst, 2020, p 451). I interpret this as an argument for the necessity of taking power relations seriously.

Arctic governance and transnational issues of justice

By mapping the most influential transnational actors in the region of the Arctic, it becomes clear that there are important international, regional, national and local dimensions at play and they each influence issues of justice in different ways. The Arctic is a region where several different nation-states and peoples share a connection and often share interests. The power dimensions in the region are, therefore, interwoven between states, state governments, transnational corporations, local businesses, and local and Indigenous communities. A complex web of stakeholders and actors emerges as a result and thus justice must account for this complexity. The fact that the Arctic is a region where eight different nation-states have sovereignty over different parts of the territory provides a clear international dimension to the region. Canada, Denmark, Finland, Iceland, Norway, Russia, Sweden and the United States all have parts of their territory in what is seen as the Arctic. Moreover, several Indigenous and minority groups reside within and across these states in the Arctic area. I argue that this is in itself an argument for a theory of justice that is not state-centric, but rather flexible so that it accounts for different forms of governance. Here, Forst's theory of transnational justice seems plausible and helpful for making sense of some of the issues of justice and injustice in the region.

The Arctic is also a region where non-Arctic states and organizations (such as the European Union) have interests and stakes, partly due to still-disputed sea jurisdictions (where the boundaries between national and international waters are debated), but also due to the potential natural resources in the region, as well as increased access to trade routes that open up as the ice melts. This creates asymmetries both between Arctic and non-Arctic states, but also within the group of Arctic states, as they have a variety of land that is understood as Arctic territory. However, the asymmetry between Arctic and non-Arctic states seems to be reasonable, particularly based on the principle of sovereignty, but not necessarily by the 'all affected' principle. These principles have different scopes, as the first is centred on international relations, whilst the second goes beyond a state-centric perspective. I will return to a discussion of these later.

Traditionally, a state-based perspective has been the dominant way of making sense of international relations as well as global justice. Realist reasoning, which argues that power is the most central force for action between states, has often driven the development of international relations theory. A transnationalist perspective, by challenging a realist international relations perspective while still taking states as central players, would allow taking the state-centric level seriously, but also expand on the agency of other influential actors. Forst argues that the first aspect to address when thinking about 'issues of justice that transcend the normative boundaries of states is whether one is looking for principles of international or of global justice', and he argues for a conception of transnational justice that provides an alternative to both globalist and statist views (Forst, 2001, p 120). The statist view focuses primarily on states, while a globalist view tends to focus on persons as subjects of justice. This implies that people and peoples are the units of analysis parallel to nation-states for a transnationalist theory of justice, while person refers to the individual. Forst's theory is based on a universal, individual right to reciprocal and general justification, and he argues for justifiable social and political relations both within and between states (Forst, 2001, p 120). Given that the Arctic is a region with the previously stated eight-state international composition, as well as a region where several minority and Indigenous groups are residing, it seems plausible to assess issues of justice from a transnational point of departure.

Let us return to the discussion of governance in the Arctic. One way of addressing and facilitating governance of the region was the creation of the Arctic Council, established by the 1996 Ottawa Declaration, which has proven to be a crucial platform for collaboration, coordination and interaction between the Arctic states and peoples. Scholars argue that the Council has achieved considerable success in identifying emerging issues in the region and transforming them into policy considerations (Kankaanpää and Young, 2012). However, taking states as a starting point for issues of interaction indicates a national or international state-centric perspective, which seems to be a common starting point when addressing different issues in the region. There are, commonly, issues focusing on political arrangements, tensions, conflict and collaboration between states. However, an important factor in creating the Arctic Council and a unique aspect of its structure is the status of permanent participants that six Indigenous groups have. The political solution of creating the Council, therefore, moves beyond the state-centric approach, allowing for non-state groups of central importance in the region to have a say in the debates taking place on the Council. The Aleut International Association, the Arctic Athabaskan Council, the Gwich'in Council International, the Inuit Circumpolar Council, the Russian Association of Indigenous Peoples of the North and the Sámi Council have the status as permanent participants which grants them some kind of participation in discussions that play an important role in forming the policy agenda of the Arctic. This novel way of organizing regional governance structures, combining international and transnational perspectives, offers new insights when it comes to questions of participation, representation and recognition – concepts that are central to understanding justice and injustice in the Arctic.

One of the primary reasons that the considerably broad representation at the Arctic Council is important is that it could have significant trickle-down effects on both international and national politics and policies. It is also crucial to acknowledge, particularly from a justice perspective, that actors other than states are recognized as formal members. This is important given that the 'all affected' principle is respected to a larger degree than in a pure state-centric organization. Further, it could be interpreted that a transnationalist approach to justice is already present in the Arctic, at least in Arctic governance structures by the formation of the Arctic Council. Importantly, though, acknowledging this as a crucial step for justice in the region is far from saying that the Arctic is a region governed justly. Here, it makes sense to differentiate between procedural justice and justice as recognition. The procedural setup seems to be more just than several other primarily state-based organizations. However, whether that is upheld throughout debates, discussions and decision making is an empirical question beyond the scope of this chapter.

Arctic exceptionalism?

One aspect alluded to earlier is the one of Arctic exceptionalism. This is a theme debated in Arctic scholarly discourses, primarily when it comes to international relations and security studies, even though it has been argued that this narrative is insufficient for understanding the complex security situation in the region (Hoogensen Gjørv and Hodgson, 2019). Questions in this debate are concerned about whether the Arctic is special and/or different from other regions. What is of primary relevance for this chapter and volume is whether or not the Arctic is *that* different when it comes to issues of justice (rather than security) – and why justice, as a critical notion, could be helpful for better understanding the societal, territorial and environmental aspects of the Arctic. This is driven by the questions of why and how the notion of justice could be understood in the Arctic. This set of questions governs this chapter, yet the Forstian perspective of transnational justice is at the centre of the analysis. I would argue that this transnational perspective has a generalist grounding, as Forst's principles of generality and reciprocity are developed with a universalist approach in mind, yet allowing for context sensitivity. These principles could potentially challenge the argument of Arctic exceptionalism.

A common argument in regional research is the one of particularity, or exceptionalism, in that the region under study is exceptional, different or deviating in several ways. This kind of argument is common also for research in the Arctic (Käpylä and Mikkola, 2019; Hoogensen Gjørv and Hodgson, 2019). Some of the arguments for this Arctic exceptionalism build on factors such as the political setup, but also issues of climate and environmental challenges. The Arctic is the most heavily affected region by climate change in the world, as the average temperature rise has been shown to be higher and much more rapid in the Arctic than in other parts of the world (Vincent, 2020). Further, this rapid change heavily affects vulnerable and exposed populations which are often minorities and Indigenous peoples primarily living in remote regions who maintain strong links to the environment through their livelihoods (Coggins et al, 2021; Mattar et al, 2020). The developments related to climate change and the precarious situation of the Arctic have, in tandem with the increased economic activities in the Arctic, contributed to the issues of justice becoming increasingly important in the region. It seems to speak to the narrative of Arctic exceptionalism. However, it also seems clear that what happens in the Arctic affects the planet, due to issues like climate change. This adds to the argument of the 'all affected' principle, but if people in other parts of the world are affected by what happens in the Arctic, should they also have a say in debates in, for instance, the Arctic Council? This seems to be a politically difficult solution, as it challenges the principle of sovereignty, yet it speaks to the core of representation. Is it then reasonable to argue that, even if the Arctic Council provides a favourable setup for Indigenous groups, it is still lacking when it comes to full representation? It seems that this argument can be made when considering the 'all affected' principle. Following the logic of the principle of sovereignty, however, it is not so clear.

There seem to be at least two risks to adopting accounts of justice in the Arctic based on Arctic exceptionalism. The first one is the risk of limiting our understanding of the Arctic, while the second is the risk of limiting our understanding of justice. Therefore, it seems important to allow for theoretical flexibility, both in the conceptualization of the Arctic and in the conceptualization of justice in the Arctic. I argue that Arctic exceptionalism seems not to be theoretically reasonable and that this is more of an empirical question. However, in this chapter one account of justice is assessed for the region. This is not to state that other accounts of justice are not worth exploring, but rather to add to the complex puzzle of justice in the Arctic by exploring one corner of the puzzle.

Assessing a Forstian transnational theory of justice in the Arctic

What happens when we take Forst's theory of justification and transnational injustice seriously while analysing the general conditions of governance in the Arctic? First, based on the reasoning noted earlier, it becomes clear that the theory of a right to justification is a cosmopolitan one, understanding people as having the same status regardless of where they are citizens. This could be a challenge when it comes to justice in the Arctic Council, if we consider the unit for analysis is not primarily persons, making the Council embedded in a statist-dominated structure.

Let us return to the principles mentioned earlier by beginning with the *principle of sovereignty* and the *'all affected' principle*, in relation to the Arctic. First, the principle of sovereignty could be seen as one approach to distributing power and voice between states, and as all Arctic states are recognized as sovereign parties with equal legal standing in international law, they have equal standing. This clearly has an international, state-based dimension. The 'all affected' principle, on the other hand, is not limited to nation-state borders but is instead rather transnational in character. In the context of the Arctic, and in light of discussions in climate justice debates, the effects of what is happening in the Arctic affect people across the globe. This would imply that the 'all affected' principle has a potentially wider scope when it comes to issues of sustainability and justice in the Arctic. But would that also imply that all affected by what is happening in the Arctic should have a say in Arctic governance? I would argue that the 'all affected' principle needs some limitations based on a proportionality assessment. It does not seem to be reasonable – or at least not realistic – for everyone in the global community to have the opportunity to participate in negotiating local solutions in the Arctic. This offers a parallel to Berit Skorstad's discussion in Chapter 7, as this is also relevant for climate policy: is it reasonable or just that the global need for copper should trump the need to protect Arctic nature?

What seems important when it comes to transnational justice in the Arctic Council is that arguments are delivered in a reciprocal and general way. Meaning that any state or group representative must make claims that their own leaders and fellow representatives would allow in their own country or territory. You cannot make claims to others you would not accept yourself. Further, if all 'those for whom the norms are valid', then all of the members – regardless of their status – should be equally involved. This seems to indicate that the formal aspects of generality are accounted for when it comes to generality in the Arctic Council, but we need an empirical study to assess the viability of reciprocity on the Council.

Conclusion

This chapter has discussed different aspects of justice and it has stressed the need to pursue different types of justice theorizing in order to better understand what justice and injustice are in the Arctic. The reasoning has centred on issues of justice and injustice on a transnational level, even though the arguments are of general character. One of the takeaways of this chapter is that it is crucial to be sensitive to regional and local circumstances of justice and injustice. This is the case even though the chapter ends in a position favouring an application of general theories of justice in the context of the Arctic, utilizing the normative principles of reciprocity and generality, yet allowing for contextual differences. This further implies a humble scepticism towards the narrative of Arctic exceptionalism.

The chapter has been governed by a few questions. Firstly, I asked if it is reasonable to understand the Arctic as special or different from other regions when it comes to issues of justice. What has been argued throughout the chapter is that issues of justice in the Arctic are central to the peoples of the Arctic, but also to some extent to a global audience, as events in the Arctic affect people across the globe due to climate change. This makes a discussion of who is affected a central one. A second question was how and why justice, as a critical notion, could be helpful for better understanding the societal, territorial and environmental aspects of the Arctic? This seems to be the case, and a Forstian notion of transnational justice has been helpful but needs further research. Certainly, justice needs to be addressed and assessed from its whole breadth of perspectives.

The third question, which covers two aspects, asks how and why the notion of justice can be understood in the Arctic. Throughout the chapter, I have demonstrated that a transnationalist account of justice seems plausible and applicable to the Arctic region. However, this has also brought a few new questions, as the focus in this chapter has been on an international and transnational level, primarily focusing on governance structures in the Arctic. I used the Arctic Council as an example, which arguably has proven to be useful for thinking about transnational justice in practice. As stated earlier, numerous different aspects could be raised when it comes to issues of justice, or injustices, in the Arctic. These could be seen as located on different levels, yet often connecting more than one level to another. Novel accounts in previous research, including several chapters in this volume, offer innovative insights on different aspects of justice – or rather injustices – from various sectors across the Arctic. For instance, reindeer herders' struggles for access to land, the placement of windmills in traditional land areas or the unevenly distributed consequences of climate change that present severe and concrete challenges for several people and peoples in the Arctic. These examples include issues of what is the right or wrong thing to do, but also to what extent the approaches or actions trying to address these are just. This indicates that there are moral aspects intertwined into the issues of justice, which are clearly also of legal and political character. Identifying these moral aspects is crucial and strengthens a normative approach if the aim is to discuss the issues through the lens of ethics.

However, seeing issues of justice and injustice in the Arctic as only moral problems (issues of right and wrong) risks providing a simplified understanding, even though elements of right and wrong are necessary for understanding issues of justice and injustice. As this volume shows, these issues are much more complex, intertwined and multi-layered, and commonly have moral implications. Justice theorists have shown that an action, which is seen to be right, could be unjust, while an action that seems wrong could be just, depending on perspective or starting point. This is largely a matter of which perspective is taken as a starting point, which proves that paying attention to different aspects of justice is crucial for better understanding the potential justness of developments in the Arctic.

Study questions

1. What are the pros and cons of focusing on justice versus injustice?
2. What are the strengths and weaknesses of a statist, globalist and transnationalist account concerning justice?

3. What are the strengths and weaknesses of a transnational approach to justice in the Arctic?

Acknowledgements

This chapter has received funding from the European Union's Horizon 2020 research and innovation programme under grant agreement No 869327.

The author wishes to thank fellow contributors, in particular Berit Skorstad and Aaron Cooper, for helpful comments in revising this chapter, and Michael Shirley for language and copyedits.

References

Coggins, S., J.D. Ford, L. Berrang-Ford, S. Harper, K. Hyams, J. Paavola, I. Arotoma-Rojas, and P. Satyal (2021) 'Indigenous peoples and climate justice in the Arctic', *Georgetown Journal of International Affairs*, February. Available from: https://gjia.georgetown.edu/2021/02/23/indigenous-peoples-and-climate-justice-in-the-arctic/ [Accessed 25 October 2021].

Crawford, B.K. (2021) 'Explaining Arctic peace: a human heritage perspective', *International Relations*, 35(3): 469–88. https://doi.org/10.1177/00471178211036782.

Forst, R. (2001) 'Towards a critical theory of transnational justice', *Metaphilosophy*, 32: 160–79. https://doi.org/10.1111/1467-9973.00180.

Forst, R. (2014) *The Right to Justification: Elements of a Constructivist Theory of Justice*, New York: Columbia University Press.

Forst, R. (2016) 'The justification of basic rights: a discourse-theoretical approach', *Netherlands Journal of Legal Philosophy*, 3: 7–28.

Forst, R. (2017) *Normativity and Power: Analyzing Social Orders of Justification*, Oxford: Oxford University Press.

Forst, R. (2020) 'A critical theory of transnational (in-)justice: realistic in the right way', in T. Brooks (ed.) *The Oxford Handbook of Global Justice*, Oxford: Oxford University Press, pp 451–72.

Heininen, L. (2018) 'Arctic geopolitics from classical to critical approach – importance of immaterial factors', *Geography, Environment, Sustainability*, 11(1): 171–86. https://doi.org/10.24057/2071-9388-2018-11-1-171-186

Hoogensen Gjørv, G., and K.K. Hodgson (2019) '"Arctic exceptionalism" or "comprehensive security"? Understanding security in the Arctic', in L. Heininen, H. Exner-Pirot and J. Barnes (eds) *The Arctic Yearbook 2019: Redefining Arctic Security*, Akureyri, Iceland: Arctic Portal, pp 218–30.

Kankaanpää, P., and O.R. Young (2012) 'The effectiveness of the Arctic Council', *Polar Research*, 31. https://doi.org/10.3402/polar.v31i0.17176

Käpylä, J., and H. Mikkola (2019) 'Contemporary Arctic meets world politics: rethinking Arctic exceptionalism in the Age of Uncertainty', in M. Finger and L. Heininen (eds) *The Global Arctic Handbook,* Cham: Springer, pp 153–69. https://doi.org/10.1007/978-3-319-91995-9_10

Keil, K. (2014) 'The Arctic: a new region of conflict? The case of oil and gas', *Cooperation and Conflict,* 49(2): 162–90.

Kortetmäki, T. (2018) 'Can species have capabilities, and what if they can?', *Journal of Agricultural and Environmental Ethics,* 31: 307–23. https://doi.org/10.1007/s10806-018-9726-7.

Mattar, S.D., M. Mikulewicz, and D. McCauley (2020) 'Climate justice in the Arctic: a critical and interdisciplinary climate research agenda', in L. Heininen, H. Exner-Pirot and J. Barnes (eds) *The Arctic Yearbook 2020: Climate Change and the Arctic: Global Origins, Regional Responsibilities?,* Akureyri, Iceland: Arctic Portal, pp 260–85.

Pincus, R., and the Foreign Policy Research Institute (2021, April 6) *Braving the Cold: Climate, Competition, and Collaboration in the Arctic Sea Lanes* [video] YouTube. Available from: https://www.youtube.com/watch?v=kYqbxVnq6I4 [Accessed 25 October 2021].

Rauschmayer, F., T. Bauler, and N. Schäpke (2015) 'Towards a thick understanding of sustainability transitions – linking transition management, capabilities and social practices', *Ecological Economics,* 109: 211–21.

SeaPower (2021, January 28) 'U.S. coast guard, Russian border guard patrolled maritime boundary line', [online], Available from: https://seapowermagazine.org/u-s-coast-guard-russian-border-guard-patrolled-maritime-boundary-line/ [Accessed 25 October 2021].

Sidortsov, R., and A. Badyina (2023, in this volume) 'Expanding collective capabilities to conceptualize and assess the impact of oil and gas activities on the energy transition in the Arctic', in C. Wood-Donnelly and J. Ohlsson (eds) *Arctic Justice: Environment, Society and Governance.* Bristol: Bristol University Press.

Similä, J., and M. Jokinen (2018) 'Governing conflicts between mining and tourism in the Arctic', *Arctic Review,* 9: 148–73. https://doi.org/10.23865/arctic.v9.1068.

Tunsjø, Ø. (2020) 'The great hype: false visions of conflict and opportunity in the Arctic', *Survival,* 62(5): 139–56. doi: 10.1080/00396338.2020.1819649.

Vincent, W.F. (2020) 'Arctic climate change: local impacts, global consequences, and policy implications', in K. Coates and C. Holroydc (eds) *The Palgrave Handbook of Arctic Policy and Politics,* London: Palgrave Macmillan, pp 507–26.

Willand, N., and R. Horne (2018) '"They are grinding us into the ground" – the lived experience of (in)energy justice amongst low-income older households', *Applied Energy,* 226: 61–70.

2

Responsibility of and for Structural (In)Justice in Arctic Governance

Corine Wood-Donnelly

Introduction

The Arctic is a political landscape in development, and it is subject to multiple and often competing claims of sovereignty. Although situated at the margins of territorial governance of the Arctic states until recent decades, the region has experienced rapid transformations, not least in its governance arrangements. The region continues to be perceived as a zone for economic development, and meanwhile it has been identified as ground zero for global climate change. In this, the Arctic is defined as a material landscape and frameworks of sovereign property rights smooth its integration into the global economy. Its political landscape is coupled with the material landscape and the exercise of authority over decision making for the region through its governance structure is notable for power asymmetries. Focusing on core features of rules, interests and agents from the International Relations theory of Social Constructivism, this chapter interrogates the asymmetric relationship between states, Indigenous groups and non-Arctic states in the context of governance via claims to sovereignty through Iris Marion Young's (IMY) four features of social-structural processes and the five faces of oppression: exploitation, marginalization, powerlessness, cultural imperialism and violence (1990). It will also discuss the responsibility for structural justice within the structure of Arctic governance, with specific reference to the Arctic Council.

Notions of structural justice first emerge in John Rawls's veil of ignorance and the premise of fair relations needed for social cooperation in the social contract between citizen and state for an ideal structural justice to exist (Rawls, 1971). This perspective is common across conversations of structural (in)justice where the impacts and differences of relative social position result in consequential and often negative effects as a result of membership in a particular social group (Powers and Faden, 2019; McKeown, 2021). While relative positions and even the structures themselves can change or evolve, legacies of the differentials of power and advantage have an enduring impact on social capabilities (Nussbaum, 2013). This has been frequently discussed in domestic analyses of structural injustice, but it is IMY that first analysed structural injustice as a product of global and transboundary impacts resulting from unjust structural arrangements (Powers and Faden, 2019; McKeown, 2021). Structural injustice is found within the governance structures which are shaped by the repetition of processes established through accepted norms and the co-constituted rules that elevate the interests and preferences of agents with power.

Structural injustice exists when 'processes enable others to dominate or to have a wide range of capabilities available to [them]' (Young, 2011, p 52). It is caused by social processes that put groups of people 'under systemic threat of domination or deprivation of the means to develop and exercise their capabilities' (Young, 2011, p 52) and is directly 'attributable to the specific actions and policies of states or other powerful institutions'. (Young, 2011, p 45). It largely takes place 'within the limits of accepted rules and norms' and simply 'as a consequence of many individuals and institutions acting to pursue their particular goals and interests' rather than as a consequence of purposeful, targeted harm (Young, 2011, p 52). The evidence for structural injustice can be found in relational inequality 'where the more powerful agents, in following their preferences, discount the weight of legitimate claims of the less powerful agents' (Heilinger, 2021, p 187). The results of structural injustice are 'the disempowerment of members of particular social groups' by 'systematically thwart[ing] their access to resources, opportunities, offices and social positions normally available to other groups' (Ypi, 2017, p 9).

Constructing the structure of injustice

The first feature of IMY's taxonomy of structural injustice posits that social-structural processes are experienced objectively and can be both enabling and constraining within macro-social spaces (Young, 2011). This is manifest through a variety of features including 'legal rules, social norms and the material world' (McKeown, 2021, p 3) where agents behave as though the structure is real. International Arctic governance institutions, though demonstrating some innovation, follow the norm of the international system that places the sovereign state at the apex of power hierarchies and seeks to legitimize their authority as decision makers for determining who can benefit from the privileges and opportunities within this geographical space. Structures of governance are inherently established to maintain rules and norms, are based on normative positions reflecting the interests of those creating the structure, and, ultimately, determine who participates and has influence in decision-making processes.

In framing a discussion of structural injustice for Arctic governance, it is important to take one step back to look at the rules, norms and processes that have influenced the superstructure of the international system in which the meso-level of Arctic governance is situated. This structure has been described as a 'culture' with the structure 'organised by the shared understandings governing organized violence' (Wendt, 1999, p 313). Its structure is something that 'exists, has effects, and evolves only because of agents in their practices' (Wendt, 1999, p 185). The primary agent of the superstructure is the sovereign state, which maintains a monopoly on authority, power and violence to ensure its survival. This survival is also dependent on international recognition to establish the legitimacy of that monopoly. Critical to understanding structures within International Relations, for both the superstructure and meso-level governance, is in realizing their intersubjectivity – where actions are based on meaning and meaning results from interactions (Zehfuss, 2002).

Contemporary Arctic governance has developed in the post-Cold War phase of the international system. This system features a plethora of layered rules, both tacit and codified, that guide the expected behaviour between states as they engage in international relationships. Yet these rules have an older history and are deeply embedded in the establishment of the international system, first through norms of imperialism and colonization causing the dispossession and oppression of peoples and territories around the world, including the Arctic. The 20th century saw a shift away from classical imperialism and the rise of local self-determination; however, this resulted in neoliberal imperialism that, although more subtle, continued to repeat patterns of domination, including asymmetric power and economic relationships (Wood-Donnelly, 2014), socio-processes also described in IMY's global connection model, and responsibility for justice.

An essential understanding of the structure of the international system is the rule of sovereignty, which has both internal and external characteristics. In the internal realms and over its citizens and specific territories, 'the sovereign state monopolizes the violent power' (Biersteker and Weber, 1996, p 190), creating order that makes up the glue of the social contract. In its external realm, where relations exist between states, sovereignty is the recognition of that monopoly of power by other states. In this recognition 'states extend to one another the system of mutual recognition that creates a society of states, reflecting and embodying state supremacy' (Biersteker and Weber, 1996, p 190). Sovereignty is a rule that is 'negotiated out of interactions with intersubjectively identifiable communities' (Biersteker and Weber, 1996, p 11); it is the trump card of international relations.

Embedding social inequality within the structure

The second aspect of IMY's taxonomy suggests asymmetries in a relative social position within a structure create societies which place limitations on agents' actions, causing inequalities to 'thicken' and be reinforced over time. This has the effect of 'positioning people prior to their interactions and condition the expectations and possibilities of interaction' (Young, 2011, p 57). The Arctic Council is structured with a tripartite hierarchy: Members, Permanent Participants and Observers. While international Arctic governance is itself a meta intersubjective community, membership within this community reinforces inequality within the relative social positions of the participating agents, through acceptance and repetition of community norms. This hierarchy can be identified as 1) Members, 2) Permanent Participants and 3) Observers of the Arctic Council, embedding inequality as a normative operator within the governance of the region.

The Arctic Council, the foremost international Arctic governance structure, is unusual in that states are not treated equally within the hierarchy of participation by situating Indigenous groups with a higher status than non-Arctic states. Despite this elevated position, the Arctic states have in actuality reinforced the rule of structural hierarchy that posits states as the dominant agents within international relations. This inclusion of Indigenous groups within the decision-making processes of regional governance has been lauded as a great step forward for the international system; however, it is arguable that the participation of Indigenous groups within Arctic governance structures does not restore Indigenous equality nor recognize their sovereignty, but is merely a method of imperialism whereby states can legitimize their authority over the Arctic through this social cooperation. The inclusion of Indigenous groups within international governance structures does not challenge the hierarchy of agents within the international system nor does it equalize participation in decision-making processes.

Governance in the Arctic has steadily developed into a cohesive structure for managing the emerging issues of the Arctic, solidified with the establishment of the standing Secretariat for the Arctic Council in 2012. The Secretariat is intended 'to strengthen the capacity of the Arctic Council to respond to the challenges and opportunities facing the Arctic' (Arctic Council, 2011), giving the structures of Arctic governance greater formality and consistency. In the space of fewer than thirty years, governance in the region has blossomed into a mature structure where Members can collaborate to address the issues specific to the region: environmental changes, changes to the human dimensions (including effects on traditional Indigenous lifestyles), and the impacts of resource exploitation. Yet this structure elevates the interests of one group over the interests, and perhaps needs of other agents.

Legitimizing the structure through repetition

The third feature of IMY's taxonomy is about the construction and cementing of the structures, norms and processes through agents' actions, where the 'structures are co-constituted as they are created and produced through the repetition of norms and the through actions of actors'. This symbolic interactionism suggests that 'the social world is constructed through mundane acts of everyday social interaction' where through repetition 'social groups constitute symbolic and shared meanings' (Del Casin and Thien, 2020, p 177). In this reiterative process, the 'rules and resources that define structures exist only insofar as the individuals in the society have knowledge of them, see them as creating possibilities for themselves, and mobilize them in their interactions with others' (Young, 2011, p 60).

Rules, which may be codified or merely social norms, are intersubjective understandings 'that tells people *what* we *should* do' (Onuf, 1998, p 59) and act as a limiter to the options and potentials of interaction. They are established to 'shape normative and ideological frameworks that constitute stable patterns of interaction' (Burch, 2000, p 187) and, once introduced, are legitimized through repetition and amplification across a social network. Transformation in a system occurs because of the introduction of a new rule, or a new shared understanding, which influences the normative behaviour of the agents operating within that system where 'meaningful action is created by placing an action within an intersubjectively understood context' (Kratochwil, 1989, p 24).

In the Arctic, the norm of sovereignty is a powerful motivator for the actions of specific members of this society. As Kratochwil explains (1989, p 251), the concept of sovereignty was used to legitimize internal structures of hierarchy within the state, and, from this concept, the notion of legal equality between sovereign states. In systems of international governance, this norm is repeated by actors, reinforcing and legitimizing the monopoly of power of the state, both in internal and extra-territorial affairs. As participating agents repeat the relationships provided in governance structures and as non-Arctic actors clamour to be accepted as Observers, they normalize this inequality through these performances, causing the rule to 'thicken' and become more stable.

Within the structure of Arctic governance, extra-territorial space is absorbed into the sovereign control of the dominant agents, those agents are the Arctic states. Anyone who is not an 'Arctic state', operate under a different regime, which as IMY describes, 'what differentiates social positions is that different rules apply to different people in different positions' (Young, 2011, p 60). Within the international society of Arctic governance, institutions have been created in a particular context and introduced into 'a "regulatory" space already occupied by a set of problem definitions and policy strategies'

(Hanf and Underdal, 1998, p 161). The result is a continuous layering of institutions, regimes and other normative expectations, which together construct an international system understood through 'shared knowledge, material resources, and practices' (Wendt, 1995, p 73).

Processes and consequences of structural injustice

The fourth feature of IMY's typology of injustice positions that the processes that create structures have consequences, often unintended for actors within the structure based on their power to influence the shape of the structure leading to 'vulnerability to deprivation and oppression for the least advantaged agents in the structure' (McKeown, 2021, p 3). The Arctic as a region in need of political organization has been accepted by the states with national interests in the Arctic region, who as a result are cooperating to protect and expand on these interests, ranging from the sovereignty of territory to exploitation of transboundary resources.

The international system in which Arctic governance is constructed is not a tangible structure. Rather, it is a product of shared understandings between agents who accept that the structure does exist. Like all systems, the international system is ordered by certain rules, principles and procedures but it is a 'social structure that exists only in process' (Wendt, 1995, p 74). However, this particular result for the least-advantaged agents is not novel to the international system, rather it is the perpetuation of historical injustices created in an older iteration of the international structure. This refers to an international system ruled by imperial practices of territorial land grabs and the disenfranchisement of Indigenous people from their traditional homelands, resources and self-governance.

This system, as a product of social and political development, is subject to change as new forces act upon it and as new rules are introduced to the system. The development of international Arctic governance arose out of the need for states to counteract the condition of anarchy in the international system so that issues common to states in this geographic space, such as environmental protection, could be addressed. This governance emerged as a counterbalance to international anarchy where existing international law mechanisms fail to fully address the interests of states in the region. However, by focusing on the states as the dominant agents for decision making, the process of creating this governance system nearly excluded representation of and from peoples living in the Arctic.

Indigenous peoples ultimately gained participation in Arctic governance as Permanent Participants in the Arctic Council. This emerged first in the Arctic Environmental Protection Strategy (AEPS) through the insistence and courage of the Inuit Circumpolar Council President, Mary Simon (Yefimenko, 2021). Yet, when the AEPS transformed into the Arctic

Council, this participation was not guaranteed as the new rule. In fact, in the late-stage negotiations for the creation of the Arctic Council, the role and status of Indigenous peoples were nearly downgraded to that of mere observers (Brøndbo, 2016). So, while the desire of the Indigenous citizens of the Arctic to participate in the discussions around issues of the region might be included in their interests, it is not yet in states' interest to elevate Indigenous peoples to the hierarchical level of states, which has consequences for their autonomy, self-determination and prosperity.

Representing the wider international community, Observers to the Council are another disadvantaged group within the structure of Arctic governance, albeit with different stakes in the game. Their membership is conditional, requiring their affirmation of the sovereignty of the Arctic states over the region, and by this admission, position themselves as having no legitimacy in decision making for the region. This is despite the interconnectedness of the Arctic to transnational and global challenges, such as climate change, ocean acidification and pollution; it remains to be seen if this poses consequences for the governance of these transboundary issues. While the international system may be premised on the sovereign equality of states, Arctic governance relegates this status in this context and confirms is it normatively possible for some states to be less equal than others. In time this may influence the underlying rule of sovereign equality of states elsewhere in the system.

Responsibility for injustice

Responsibility for justice is dependent upon the form of injustice and culpability for causing or perpetuating that injustice. IMY positions that structural justice comes in several forms: pure or avoidable and that 'structural processes operate across the boundaries of many nation-state jurisdictions' (Young, 2011, p 142). Pure structural injustice has no identifiable perpetrator, and the resulting injustice is the 'sum of multiple agents' nonblameworthy actions' (McKeown, 2021, p 4). This type of injustice can only be remedied through collective action to reverse the effects of multi-scalar agents operating through structural hierarchies of power because it is caused 'wholly in virtue of the features of social structure, and so irrespective of culpability' (Estlund, 2020, p 6).

The second form, avoidable structural injustice, occurs when 'powerful agents with the capacity to change unjust social structures' (McKeown, 2021, p 5) fail to make the necessary societal changes to eliminate injustices. This capacity relates to a combination of power, resources and opportunities to remove injustice. Finally, deliberative structure injustice occurs where 'agents are deliberately perpetuating unjust background conditions' (McKeown, 2021, p 5). This is ordinarily for their gain and when agents have the power to change these conditions but do not to avoid less satisfactory outcomes.

Responsibility for injustice can be seen within three different modalities, with differing degrees of obligation to remediate: instrumental causality, culpability in the production and reproduction through performances of injustice, and obligation to facilitate remediation (Estlund, 2020). In the instrumental causality for injustice, both the agents and the rule governing the agents' behaviour resulting in the flaws of the structure must be examined for perpetuating the injustice, although the injustice may be a product of historical actions. Culpability for injustice emerges through the production, reproduction and repetition of norms and rules that result in unjust conditions. When justice is revealed through normative evolution, actors become both obliged and responsible for facilitating change that remediates or removes injustice.

Structural injustice is the result of the tyranny of agents with the power to introduce, develop and normalize the rules and norms that exhibit inequality through disabling constraints, domination and oppression beyond the mere 'exercise of tyranny by the ruling group' (Young, 1990, p 39). IMY describes this oppression as having 'five faces', including exploitation, marginalization, cultural imperialism, powerlessness and violence, meanwhile describing justice as including 'the institutional conditions necessary for the development of individual capacities and collective communication and cooperation' (Young, 1990, p 39). There are several examples where these five faces emerge in Arctic governance.

Resource development is an ever-present undercurrent in Arctic governance, leading to the exploitation as the first face of oppression. This is understood to be the 'steady process of the transfer of the results of the labor of one social group to benefit another' (Young 1990, p 49). In the context of resource development, labour exploitation began when trading companies eroded the food security of Arctic peoples by incorporating their key traditional resources into the market economy and making them reliant on imported goods. This accelerated large-scale natural resource exploitation which frequently uses imported labour and accumulates profits outside of the Arctic – creating competition for already scarce resources such as housing and food. Meanwhile, these operations can degrade environmental conditions, reducing the capacity of traditional economies vulnerable to compromised ecosystems (Duhaime and Caron, 2006). This exploitation includes not only labour, but critically compromising capabilities through the material deprivation of communities and the broader dynamics of resource exploitation.

The results of this exploitation contribute to the marginalization, the second face of oppression where 'a whole category of people is expelled from useful participation' (Young, 1990, p 53), and for Indigenous peoples, this is 'politically, economically and epistemologically' (Comberti et al, 2019, p 15). Material deprivation sometimes means that Permanent Participants may not have adequate resources to enable participation in governance processes, including attending meetings. Although allowed participation by the Council charter, when Permanent Participants are in attendance, they are not fully included as decision makers in governance processes. Instead, they must rely on 'the states within which they reside [to] speak on their behalf, yet the history of marginalization and discrimination by these same states undermines the legitimacy of their representation' (Comberti et al, 2016). This marginalization is promoted by the normative expectations that the sovereignty of states, and especially of Arctic states, generates decision-making legitimacy.

Powerlessness, IMY's third face of oppression, is a relational understanding of power where 'only states have the institutional capacities to adjust patterns of advantage and the politically legitimate authority to regulate relations' (Powers and Faden, 2019, p 147), resulting in 'a pattern of unequal consequences' (Onuf, 2013, p 283). In a structure with embedded hierarchies and limited capacity to change these inequalities, both Permanent Participants and Observers of the decision-making process of Arctic governance are subjected to the consequences and effects of policies and even the agenda-setting of those holding the power. For the wider global community, this prompts questions of cosmopolitan justice emerging from the distribution of harms from resource exploitation, transboundary environmental damage and, indeed, climate change, related to IMY's social connection model of responsibility.

Cultural imperialism marks IMY's fourth face of oppression, where the culture of decision makers is installed as the normative order through the 'universalization of a dominant group's experience and culture', which promotes 'the experience, values, goals and achievement of these groups' (Young, 1990, p 193). This emerges not only in Arctic governance reaffirming the primacy of states from the culture of an international system founded on classical imperialism, but also in that it is the interests and frequently the national interests and needs of those with power which are the most widely communicated and actioned issues. This also includes subsuming the interests of environmental protection to the culture of capital accumulation and economic development and deprioritizing climate governance over resource exploitation (McCauley et al, 2022; Wood-Donnelly and Bartels, 2022).

Violence, IMY's final face of oppression, is a social practice that includes 'not only direct victimization' but also group knowledge 'shared by all members of oppressed groups that they are liable to violation' (Young, 1990, p 62). While violence is often viewed as physical, such as in acts of war, dispossession of resources or removal of children from their communities, it extends beyond this. It can also include subjugation of groups through exclusion from equal roles, or even the recognition of legitimate participation in decision making, and especially the diminishing or silencing of voices in agenda-setting areas of focus. While it is easier to reflect backwards on violence and causality for injustice in the Arctic today, it remains to be seen how these 'new' injustices will be perceived in generations to come.

Discussion

International governance is a mechanism providing a means to manage the conditions created by shared understandings in the international system addressing the problem of anarchy, which results in two deficiencies in international relations. First is that under anarchy there is a lack of an overarching global authority to order the behaviour between states. Second, is that the international system is comprised of several hundred discrete political units, each with control over a finite territorial space. Without the ordering power of political authority, the spaces beyond the sovereign borders of the state retain all the insecurity and competition that arises from a lack of political ordering. Thus, international governance provides a mechanism whereby states can overcome the anarchy in international spaces by creating authority over these extraterritorial spaces.

The development of governance in the international realm has created conditions where states can coordinate mutual pursuit of interests through cooperation in extraterritorial areas. Current Arctic structures of governance are constructed in an international system understood through 'shared knowledge, material resources, and practices' (Wendt, 1995, p 73). Governance of the region is a system that has been pieced together, sometimes described as a 'web' (Hansen-Magnusson, 2019; Exner-Pirot, 2016). It addresses the needs of the Arctic states, closely mirroring the norms and structure of the international system, formed from a layering of institutions, regimes and other normative expectations. This governance seeks not only to address the wills and interests of the Arctic states but, in addition, by banding together in the Arctic Council has strengthened their capacity to legitimize Arctic decision making.

When including Indigenous Participants within the Arctic Council, potential existed to create meaningful stage-change in the norms and expectations of international governance, especially with regards to Indigenous peoples around the world. This was identified by Oran Young, who said: 'The Council has accepted a number of indigenous peoples' organizations as Permanent Participants in its activities, a notable precedent with implications extending far beyond the Arctic' (2009, p 428). While this transformation is in itself a step forward, it does not raise Indigenous peoples to be equivalent agents of power within Arctic governance, or beyond. Using the conceptual framework of IMY, the resulting inequality could merely be perceived as a new injustice.

States, as the dominant agents of political organization in the international system for centuries, have created rules that have positioned themselves advantageously. Some of the rules that have made states the dominant agents of international relations include those giving states a monopoly on war, sovereignty over territory and jurisdiction over their citizens, among others. These rules have been tacit, demonstrated in the behaviour of states, and sometimes explicit, such as in the instances of ceremonies of possession, which made imperial power the new overlords of distant lands, and in the creation of codified international law. States can create rules in the international system. Individuals, and even groups of individuals, do not have this ability.

Rules of territorial acquisition previously justified the annexation of the territory of Indigenous peoples of the Arctic into the sovereign domains of the circumpolar states. Although the international system now incorporates principles such as the rights to self-determination of peoples, or the innocence of non-combatants in wartime, the introduction of Indigenous representatives into the governance mechanisms does remove all inequality. The state remains the primary agent in the international system and is not yet ready for the introduction of the rule of 'Indigenous groups are equal with states', as it would upset the relative hierarchy and order of this system. Thus, it can be determined that the inclusion of Indigenous groups into Arctic governance does not create conditions for structural justice and opens up additional questions related to procedural, distributional and recognitional justice.

Conclusion

The Arctic Council, along with other elements of Arctic governance, together form an umbrella mechanism where the Arctic states can cooperate on overlapping issue areas affecting the region. This cooperation between Arctic states began through environmental protection strategies but has advanced to address Arctic-specific issues from search and rescue to scientific cooperation. Arctic governance is an intersubjective structure that has been created by the interaction of authority and social practices through mutual state recognition of the legitimacy of the monopoly of power and extension of popular sovereignty over Arctic peoples and territories. This cooperation and recognition of sovereignty within Arctic governance is an arrangement that has made it possible for the Arctic region to remain a dream space for economic development. Yet, within this structure and the rules, agents and interests it serves exists a critical flaw: this flaw is injustice.

Arctic governance, and in particular the Arctic Council, is the most progressive structure of international governance in existence today. The Council is the first international arena to recognize Indigenous groups as Permanent Participants in the discussions that underpin governance processes. However, despite this step forward, this new framework for governance both replicates old injustices and creates new injustices by elevating national injustices to the international level and by not creating conditions for full participation. Moreover, this leap forward was the result of Indigenous leaders fighting for recognition of their right to have a voice in Arctic decision making by their long-standing domicile in the Arctic region and not at the initiation of the Arctic states. Additionally, through the establishment of Arctic governance and patterns of accepted practice in the Arctic, inequality is being solidified, which guarantees that only Arctic states will ultimately make the decisions. Non-Arctic states are also excluded – even though many issues the Arctic Council addresses are transnational and transboundary problems.

Arctic governance has been developed not to address historical injustices, but to create a mechanism for Arctic states to extend their influence over issues outside of their sovereign borders in the name of their national interests. While the legal agreement of the Arctic Council facilitates norms for issues and areas external to sovereign borders, the working groups of the Council provide for transnational information sharing on common, and often transboundary, issues. With the creation of a new institution of governance that perpetuates old injustices, it may be considered that states that interact with the social processes of this structure are culpable and liable for the resulting marginalization and oppression.

The responsibility for structural injustice rests with all actors in Arctic governance. It rests with those who are culpable for causing or perpetuating the injustice. It rests with those who have the power to change the injustice but are not doing so. It also rests with those willing to perform and reinforce these unjust conditions. Those watching, observing and participating in Arctic governance should perhaps remain uneasy so long as the rules and norms of this structure perpetuate inequalities that exploit, marginalize and dominate the plethora of voices that should be heard and actioned in just and meaningful ways.

Study questions

1. What is the ideal structure of Arctic governance that would enable structural justice?
2. Who is responsible for injustices in Arctic structures and how should they be removed?

3. What does IMY's concept of structural justice, five faces of oppression and responsibility for justice contribute to International Relations theory?

Acknowledgements

This chapter has received funding from the European Union's Horizon 2020 research and innovation programme under grant agreement No 869327.

References

Arctic Council (2011) 'Nuuk Declaration', Arctic Council, Available from: http://hdl.handle.net/11374/92 [Accessed 1 June 2021].

Biersteker, T.J., and C. Weber (1996) 'The social construction of state sovereignty', in T.J Biersteker and C. Weber (eds) *State Sovereignty as a Social Construct*, Cambridge: Cambridge University Press, pp 1–21.

Brøndbo, S. (2016) 'Interview with Mary Simon', *Shared Voices*, special volume, pp 16–17. Available from: https://old.uarctic.org/shared-voices/shared-voices-magazine-2016-special-issue/ [Accessed 1 June 2021].

Burch, K. (2000) 'Changing the rules: reconceiving change in the Westphalian system', *International Studies Review*, 2(2): 181–210. doi:10.1111/1521-9488.00209.

Comberti, C., T.F. Thornton, and M. Korodimou (2016) 'Addressing Indigenous peoples' marginalisation at international climate negotiations: adaptation and resilience at the margins', *SSRN Electronic Journal* [Preprint]. doi:10.2139/ssrn.2870412.

Comberti, C., T.F. Thornton, M. Korodimou, M. Shea, and K.O. Riamit (2019) 'Adaptation and resilience at the margins: addressing Indigenous peoples' marginalization at inteikrnational climate negotiations', *Environment: Science and Policy for Sustainable Development*, 61(2): 14–30. doi:10.1080/00139157.2019.1564213.

Del Casino, V.J., and D. Thien (2020) 'Symbolic interactionism', in A. Kobayashi (ed.) *International Encyclopedia of Human Geography*, Amsterdam: Elsevier, pp 177–81. doi: 10.1016/B978-0-08-102295-5.10716-4.

Duhaime, G., and A. Caron (2006) 'The economy of the Circumpolar North', in S. Glomsrød and S. Aslaskser (eds) *The Economy of the North*, Oslo: Statistiske analyser/Statistisk Sentralbyrå, 84, pp 17–25.

Estlund, D. (2020) 'What's unjust about structural injustice?' [online], Available from: https://www.law.nyu.edu/sites/default/files/ESTLUND-DRAFT-Structural%20Injustice-NYU%20version%20Nov%202020_0.pdf [Accessed 15 October 2021].

Exner-Pirot, H. (2016) 'Open Canada', *Why Governance of the North Needs to Go beyond the Arctic Council*, 14 October, [online], Available from: https://opencanada.org/why-governance-north-needs-go-beyond-arctic-council/ [Accessed 2 February 2021].

Hanf, K., and A. Underdal (1998) 'Domesticating international commitments: linking national and international decision-making', in A. Underdal (ed) *The Politics of International Environmental Management*, Dordrecht: Springer Netherlands (European Science Foundation), pp 149–170. doi:10.1007/978-94-011-4946-4_8.

Hansen-Magnusson, H. (2019) 'The web of responsibility in and for the Arctic', *Cambridge Review of International Affairs*, 32(2): 132–58. doi:10.1080/09557571.2019.1573805.

Heilinger, J. (2021) 'Individual responsibility and global structural injustice: toward an ethos of cosmopolitan responsibility', *Journal of Social Philosophy*, 52(2): 185–200. doi:10.1111/josp.12398.

Kratochwil, F.V. (1989) *Rules, Norms, and Decisions: On the Conditions of Practical and Legal Reasoning in International Relations and Domestic Affairs* (1st edn), Cambridge: Cambridge University Press. doi:10.1017/CBO9780511559044.

McCauley, D., K.A. Pettigrew, M.M. Bennett, I. Todd, and C. Wood-Donnelly (2022) 'Which states will lead a just transition for the Arctic? A DeePeR analysis of global data on Arctic states and formal observer states', *Global Environmental Change*, 73: 102480. doi: 10.1016/j.gloenvcha.2022.102480.

McKeown, M. (2021) 'Structural injustice', *Philosophy Compass*, 16(7): e12757. doi:10.1111/phc3.12757.

Nussbaum, M. (2013) *Creating Capabilities: The Human Development Approach*, Cambridge, MA: Harvard University Press.

Onuf, N.G. (1998) 'Everyday ethics in international relations', *Millennium: Journal of International Studies*, 27(3): 669–93. doi:10.1177/03058298980270030401.

Onuf, N.G. (2013) *World of Our Making: Rules and Rule in Social Theory and International Relations*, London and New York: Routledge.

Powers, M., and R. Faden (2019) *Structural Injustice: Power, Advantage, and Human Rights*, New York: Oxford University Press.

Rawls, J. (1971) *A Theory of Justice*, Cambridge, MA: Harvard University Press.

Wendt, A. (1995) 'Constructing international politics', *International Security*, 20(1): 71–81.

Wendt, A. (1999) *A Social Theory of International Politics*, Cambridge: Cambridge University Press.

Wood-Donnelly, C. (2014) 'Constructing Arctic sovereignty: rules, policy and governance 1494–2013', Brunel University Research Archive, London: Brunel University London.

Wood-Donnelly, C., and M.P. Bartels (2022) 'Science diplomacy in the Arctic: contributions of the USGS to policy discourse and impact on governance', *Polar Record*, 58. doi: 10.1017/S0032247422000134.

Yefimenko, A. (2021) 'How Arctic Indigenous peoples negotiated a seat at the table', Arctic Council, [online], Available from: https://arctic-council.org/news/a-seat-at-the-table-how-arctic-indigenous-peoples-negotiated-their-permanent-participant-status/ [Accessed 1 November 2021].

Young, I.M. (1990) *Justice and the Politics of Difference*, Princeton, NJ: Princeton University Press.

Young, I.M. (2011) *Responsibility for Justice*, New York: Oxford University Press.

Young, O. (2009) 'The Arctic in play: governance in a time of rapid change', *The International Journal of Marine and Coastal Law*, 24(2): 423–42. doi:10.1163/157180809X421833.

Ypi, L. (2017) 'Structural injustice and the place of attachment', *Journal of Practical Ethics*, 5(1): 1–21.

Zehfuss, M. (2002) *Constructivism in International Relations: The Politics of Reality*, Cambridge: Cambridge University Press.

3

A Relational View of Responsibility for Climate Change Effects on the Territories and Communities of the Arctic

Tracey Skillington

Introduction

This chapter considers what might be deemed relevant normative standards when taking responsibility for the effects of rising global temperatures on the territories and communities of the Arctic. Are globally produced harms chiefly the responsibility of territory-specific communities in terms of their dire effects, as is often assumed? The focus will be on the unjust basis of this assumption. Alternatively, it will propose a relational model of responsibility where emphasis is placed on the interconnections between peoples, regions, climate actions and outcomes. In response to the need to actualize a more embracing conceptualization of climate justice, prospects for a 'civic connections approach' will be critically assessed, one where a cooperative imperative, when applied across sectors and regions, works to address the 'multiple domination' (Forst, 2020) experienced by climate vulnerable communities and further, seeks to establish the presumptive responsibilities of major polluters for injuries caused to the peoples, nature and landscapes of the Arctic.

A relational view of responsibility

It was Norwegian sociologist Johan Galtung (1969) who first noted how the most potent forms of violence are often those felt indirectly. The 'slow violence' (Nixon, 2013) of rising average temperatures, thawing permafrost or the spectre of charred peatlands in Siberia and their impact on local Arctic environments are undeniable today. Yet the primary sources of these harms are said to be transnational – diffuse, long-term planetary changes caused mainly by 'human industrial activity'. That is, transnationally sourced ecological, social, cultural and economic harms produced by multiple agents residing inside, between and beyond state borders. Being transnational, however, does not take from the fact that these harms constitute a distinct plane of actualization of sustained wrongdoing. By highlighting the role harm agents play in undermining adequate and healthy functioning across multiple contexts, the aim of this chapter is to show how Arctic communities come to be subject to forms of interference that not only diminish the availability of essential resources that sustain healthy patterns of living, but do so in ways that also undermine fundamental rights and freedoms (that is, harms committed without their consent, participation or best interests in mind). In this respect, it follows Philip Pettit (1996) in his understanding of how relations of domination are intricately linked to practices of sustained wrongdoing, where social, political, economic and environmental sources of harm come together to constitute, in this instance, a trans-territorial space of interference with basic freedoms, capabilities and rights. A key task of any critical inquiry into such arrangements is to explore the material environment in which multiple relations of wrongdoing are enacted and experienced as interference, for instance, interference with the self-governing capacities of a people. The peoples of the Arctic are self-governing if they preserve determinate control of important aspects of their lives together, including the capacity to establish justice in the allocation of essential resources and safeguard a shared cultural identity and common way of life. When that common life is threatened by transnationally sourced ecological devastation, the capacity to be self-governing is also threatened. Increasingly, the ecological and social circumstances of Arctic peoples' lives are rendered conditional on the plans of others to further invest in deep ocean mining, fossil fuel extraction and other carbon intensive projects (Lèbre et al, 2020). These and related activities give rise to harms that contribute to warmer temperatures, record ice melts and diminished marine flourishing in ways that disempower communities and disrupt traditional patterns of fishing, hunting, farming and mobility. In doing so, these wrongs also undermine rights to life, health, culture and security, all of which are seen as necessary for a worthwhile life but cannot be achieved by any one individual in isolation.

Such rights relate to Arctic peoples 'being' members of thriving communities and living within specific contexts where conditions are safe and plentiful. To use Amartya Sen's (1992) argument, rights must be thought of in terms of freedoms and opportunities to achieve certain outcomes important to human flourishing. For instance, the capacity of a people to remain self-determining and enjoy a safe and healthy life is dependent upon the existence of substantial opportunities to do so. Such opportunities, however, cannot be created in isolation since they are, by their very nature, relational and determined increasingly by transnational influences, a point made recently by representatives of Inuit, Sámi, Sakha, Itelmen, Yukaghir, Ulchi, Evenki, Golgan and Chickaloon communities from the Arctic, North America and Russia to the United Nations (UN) Food and Agriculture Organization at its headquarters in New York in September 2019. On this occasion the Indigenous Peoples Rome Declaration on the Arctic Region Fisheries and Environment was presented, explaining how climate change has become a major concern for all Indigenous peoples in these regions, affecting health and well-being, disrupting food chains, travel routes and hunting seasons, as well as triggering displacement in many instances.[1] Campaigners drew attention to the various ways in which the freedom to flourish is being actively undermined by unmitigated rates of global pollution, steady increases in average temperatures, more frequent wildfires, melting ice and rising sea levels, affecting adequate functioning in relation to health, well-being and security.

To ensure capability-undermining harms are connected more explicitly to the actions of specific (even if transnationally dispersed) agents, a *relational* view of climate justice is required, one that extends relations between states and regions beyond just what 'socially connects' them in their belonging together in one planetary system (Young, 2011) to a consideration of the civic connections that bind these agents to shared expectations of justice and legally grounded principles of responsibility (Skillington, 2017, pp 246–7). That is, civic connections that are legally enforceable; for instance, legal obligations to protect the capacities of all communities to remain self-determining, especially in contexts where the threat of major disaster is very real. A civic connections approach will be explored later in relation to two key concerns for Arctic regions – the growing prevalence of wildfires and ocean acidification. Although equally applicable to other issue areas (for example, accelerated loss of biodiversity or forestation), the discussion will be restricted to these two issues due to limited space. The analysis will consider how issues of responsibility are commonly defined in relation to these problems and how various alliances seek to hold territorially dispersed harm agents to account for violations of legal principles of shared responsibility and reciprocal rights and duties of care (that is, civic connections).

---

[1] See http://www.fao.org/uploads/media/FINAL_Rome_Arctic_Declaration_2019_.pdf (p 1).

Situating Arctic wildfires relationally within wider landscapes of destruction

In terms of how they are commonly presented, more frequent wildfires due to rising average temperatures (rising four times as fast in the Arctic than any other planetary region (Rantonen et al, 2022)) are widely seen as beyond the responsibility of any one identifiable group of agents of harm. By the end of July 2019, more than 745 wildfires had burned 33,200 square kilometres of land in Siberia alone; yet by the summer of 2020, the scale of damage had already surpassed that of the previous year, with more carbon produced from the beginning of January to the end of August than any other year on record. When wildfires burned through 1.5m hectares of land, family homes and forests in north-east Siberia in July 2021, locals blamed poor government preparedness and budget cuts to forestry services, but also, in particular, unusually hot temperatures linked to global climate change.[2] The general worry is that as average temperatures continue to rise, the increase in pollutants from wildfire smoke (that is, carbon monoxide, nitrogen oxides, volatile organic compounds and solid aerosol particles) will lead to further warming of the atmosphere, drier peat soils and, consequently, more fires in the years ahead (data from the Copernicus Atmosphere Monitoring Service and NASA's Moderate Resolution Imaging instruments supports this assumption).

Fires in the boreal forests and Arctic tundra, which account for 33 per cent of global land surface and hold an estimated 50 per cent of the world's carbon in soil, are expected to increase fourfold in the decades ahead due to climate change (NASA, 2019). For thousands of years, peatlands have played a key role in cooling temperatures and storing the carbon produced by accumulated organic matter. However, with rapid thawing and more intense drying, these carbon dense ecosystems are becoming more flammable, burning not just the surface of the Arctic tundra but also deep down into thick layers of carbon-rich organic matter, triggering further drying and 'legacy carbon' loss.[3] It is important to note how this discourse rarely addresses the question of responsibility from the point of view of the actions of specific wrongdoers. Instead, wildfires are typically portrayed as a problem created by 'warming temperatures' (and defined in terms of minimum relations of responsibility); that is, as fires 'that start usually by themselves' or as 'any non-structured fire other than prescribed fire'.[4] But what does it mean to define a wildfire as

---

[2] See A. Roth (2021) 'Everything is on fire': Siberia hit by unprecedented burning', *The Guardian* (20 July).

[3] NASA (2019) notes how during an intense fire when organic material containing carbon buried deep in the soil burns along with trees and plants 'legacy carbon' loss occurs.

[4] See *Collins English Dictionary* (2021).

one that 'starts by itself' in the context of climate change, especially when scientific consensus on the reasons for more frequent wildfires and primary sources of harm is clear?

Given their scale and growing intensity, wildfires triggered by rising temperatures cannot be said to be wholly 'unstructured' or 'spontaneous' occurrences. Rather, they are the product of cumulative wrongdoing, a specific side-effect of the pollution activities of extractive industries and high carbon-producing communities. It may be that fire has always been an important component of the natural order (for example, volcanoes) but as more and more of the Earth's hydrocarbons are brought to the surface and set alight, they actively contribute to a vast burnout of vulnerable regions, including the Arctic.[5] In this sense, the Anthropocene's burning landscapes are as much a product of the industrial histories, laws and policy norms of carbon capitalist regimes (large-scale deforestation, resource mismanagement, poor investment in fossil fuel alternatives) as they are a product of 'natural' processes of change. In being regulated by the political, economic, cultural and social norms of a system of minimum extra-territorial responsibility and a highly unequal global order of power (in terms of the costs and benefits of large-scale ecological destruction), they are also the product of relations of domination. The question then is whether the details of these interconnections can be specified more clearly in moral, ethical and legal terms and legitimate expectations of justice formulated?

Ocean acidification and the boundaries of responsibility

Similarly, ocean acidification is a problem that is clearly produced by specific acts of wrongdoing (regular contributions to rising CO_2 emissions levels) yet is commonly construed as governed by non-specific relations of harm and responsibility for disruptions to marine life, food chains and local economies. It is also one whose effects are felt more acutely in the oceans of the Arctic due to the fact that CO_2 dissolves faster in colder waters and in settings where 24-hour daylight makes for more active phytoplankton production and increased concentrations of hydrogen ions (Coello-Camba, 2014). Researchers forecast that most Arctic waters will lack adequate calcium carbonate minerals aragonite or calcite for shell organisms by 2080, threatening the ocean food web and fisheries of the Arctic (Katz, 2019). The Arctic Council's Monitoring and Assessment Programme (2019) for instance, estimates that the socio-economic and environmental

---

[5] For the first time in recorded history, smoke from wildfires reached the North Pole in August 2021.

consequences of ocean acidification will be substantial in the years ahead. Yet at present no serious effort is being made to impose safe limits on levels of acidification (Cassotta, 2021, p 3). The question then is why is this problem not being addressed more effectively? Perhaps one key element here is the general lack of clarity on the question of responsibility. The dominant view remains that of 'nature as object' to be carved up and divided amongst competing interests, with each responsible for the protection of their own portion. Much of the early foundations of international standards regulating the rich resources of the open seas were influenced by the claims of Hugo Grotius, who argued that the world's oceans, as common heritage, were 'unclaimable' even if sea bed resources, fish stocks and other precious properties were. The primary criterion governing access to these resources is the right of occupation. That is, the right to make use of the oceans' detachable resources for social and economic gain, or the liberty to travel through their waters.

Any 'unintended' consequences arising from interactions between resource extraction activities and already fragile marine, land and air environments in this instance are thought to be largely unassignable (unless the product of specific events such as oil spills) (Arctic Resilience Interim Report, 2013). Instead, emphasis is placed on the legitimacy of competing interests' occupation claims to the oceans. Any problems generated by such occupation, including problems of over-extraction and ocean acidification, are listed nominally as 'unintended' or 'manufactured risks' (Beck, 2006). The fact that much of this harm is inflicted on a nature that does not lend itself easily to being 'carved up' (for example, wildlife, ecosystems, migrating species, sea microbes, and so on) or divided amongst competing interests poses a problem for this model of justice, especially in terms of how responsibility for collectively produced harms might be better addressed (for example, acidification, rising temperatures) and viable long-term solutions sought for problems bigger than just the question of who owns what. Private ownership and territorial jurisdiction clearly do not exhaust entitlements to clean, pH balanced oceans or stable average temperatures.

Equally, occupancy rights cannot be assigned exclusively to essential resources, such as clean air, carbonate ions or rainfall. Instead, all earthly inhabitants share these resources as 'unclaimable' components of the commons. Even so, property rights remain the dominant element of contemporary justice reasoning. In this context, the assumption is that climate change creates bad circumstances for those who fail to take full advantage of the opportunities their own resource-rich territories create for them to adapt to deepening climate adversities. But what if those opportunities or, indeed, the capacity to avail of them are limited by devastating storm surges, rising sea levels, wildfires, the acidification of seas and so on, that is, by circumstances created by harms generated inside, between and beyond state borders? How can responsibility be allocated in this instance?

An object-centred view of natural heritage consistent with property rights is not naturally relational (that is, considerate of the needs of others). Thus, while the notion of 'shared responsibility' for the protection of natural resources and communities dependent on them may be well supported rhetorically in international discourse (for example, in the preamble to the United Nations Framework Convention on Climate Change (United Nations, 1992a)), at a deeper level it seems to run contrary to much of the reasoning of traditional political and legal thinking emphasizing a separation of powers amongst property owners and limited liability for any pollution harms generated.[6] As long as justice continues to be defined in these terms, what is right for all states to do in terms of ongoing acquisitions of limited seabed, atmospheric or land resources will not be considered in terms of what are owed to the peoples of climate vulnerable regions, including the Arctic. Instead, private gain will continue to take precedence over communal loss and insufficient attention accorded to the way agents, in their interactions with land, air and oceanic environments, affect multiple communities.

A relational view of justice remains underdeveloped due to the dominance of a property rights perspective, even though the former complies more with the normative principles and common earth reasoning embedded in several international environmental agreements, for example, the Convention on Biological Diversity (United Nations, 1992b) and the Kyoto Protocol (United Nations, 1997) (see O'Mahony and Skillington, 2012). A relational view also points in a more positive direction towards the need to protect what happens in the spaces between us by addressing how we shape each other's lives and that of a wider natural order. It also pays greater attention to the fact that what makes various natural resources 'useful', consumable or desirable in the first instance is the cumulative activities of multiple ecological agents (human and non-human). For instance, what makes a seabed rich in minerals or a soil suitable for grazing is much more than what property owners invest in them. This brings us back to the issue of how various rights, beyond mere property rights, can be brought into a more critical dialogue with the question of why we all bear responsibility for the protection of interdependent communities and ecosystems.

---

[6] With a model of limited liability, justice follows a logic of corrective rather than transformative action where compensation is allocated for individual wrongs but not larger, transnationally relevant ones.

Actualizing principles, practices and relations of co-responsibility

When defined from the perspective of those who are simultaneously recipients of universal rights and climate change wrongs (growing numbers), climate justice necessitates that violations of legitimate normative expectations of responsible action be addressed by harm doers. In these settings, the question of responsibility must connect meaningfully with the human and civic rights of overlapping communities to ensure basic capabilities, freedoms and rights are protected. Some kind of reflexive turn in the historical framework of recipience of rights, duties and responsibilities is therefore required (consolidated particularly with the establishment of United Nations in the post-Second World War period). Arguably, that turn is already being explored in some quarters.

In 2021, in advance of the Global Food Systems Summit, the Sámi Council published a declaration reminding 'UN member states, private corporations, resource centres and civil society that they must obtain our free, prior and informed consent before adopting any legislation or administrative measures or pursuing development projects and activities that may impact on our rights'. The carbon contributions of major climate offenders are noted as already contributing to grave violations of Sámi peoples' rights to life (for example, Article 2 (life) and 14 of the ECHRs), security, health, self-determination (for example, Indigenous Peoples Rome Declaration on the Arctic Region Fisheries and Environment (FAO, 2019)) and the right to democratic accountability for injuries thereby inflicted. By marking rising emissions and unregulated pollution levels as contributing to declining social and civic standards (affecting the self-determining capacities of their people), the Sámi Council reaffirms the importance of ecological heritage as a subject of social, democratic, economic and cultural, as much as ecological justice.

It also reinforces the interconnections between particular pollution activities and the growing challenges they pose in terms of the abilities of Sámi peoples to achieve healthy and democratically meaningful functioning. The Arctic Region Declaration in Preparation for the Global Food Systems Summit (2021) notes the importance of the overlap between these various forms of justice when it states: 'We reaffirm our interdependent, interrelated, interconnected and indivisible rights as elaborated in the UN Declaration on the Rights of Indigenous Peoples, including our right to self-determination or right to harvest the food we rely upon and our lands, territories and resources'. Campaigners recount how major carbon contributors, in failing to control both territorial and overseas emissions (through the export of fossil fuels, the import of embodied carbon, or the further financing of fossil fuel and mineral extractions) share responsibility for the injuries the Sámi people, as legally protected subjects, endure as a consequence of multiple wrongful acts. Since these injuries are not 'relationally justifiable' (Forst, 2017) nor produced with the welfare of the Sámi people in mind, they are said to violate all major human rights treaties and a basic principle of transparency of decision making.

What the Sámi Council highlights here are concerns also raised in the complaint lodged by sixteen youths from various world regions to the UN Committee on the Rights of the Child on 23 September 2019 against five high-polluting states (that is, Argentina, Brazil, France, Germany and Turkey). The common point being the urgency of addressing harms produced by multiple, trans-territorial climate wrongdoers and the need to bring those responsible to justice. Whilst the shift in focus towards multiple harm doers is, in many ways, a positive move in political and legal reasoning, it, nonetheless, poses considerable challenges. Efforts to highlight the transnational nature of wrongdoing simultaneously requires that campaigners prove that traditional lines of separation between peoples, state territories, communities and generations do not exonerate states from fulfilling various extra-territorial, context-transcending legal obligations (for example, duties to protect). To address such challenges, campaigners draw on a legal principle of presumptive responsibility for harms generated by multiple agents through their contributions to increasing flows of embodied carbon (leading to rising sea levels, more frequent and intense wildfires, storm surge, heatwaves, ocean acidification, and so on), prompting a need to push for a greater prioritization of responsibilities to protect the welfare of those most adversely affected.[7]

The principle of presumptive responsibility clarifies how in situations where there are a number of wrongdoers who have contributed to the generation of particular harms (for example, rising global temperatures) and where there is uncertainty as to which of them is disproportionately responsible, each wrongdoer is deemed presumptively responsible. The onus in this instance is on each wrongdoer to show how they, in fact, did not cause the relevant harms, rather than the more usual scenario where it is the injured party who carries the burden of proving harms are traceable to the actions of specific actors. Increasing legal support for this principle holds out certain possibilities for Arctic communities facing scenarios of multiple loss generated by multiple climate wrongdoers, especially as presumptive responsibility prevents states from shifting responsibility away from themselves

---

[7] We may note the relevance of the decision in the *Ilascu* v *Russia and Moldova* case (application no. 48787/99) where the European Court of Human Rights held that a 'state's responsibility may [...] be engaged on account of acts which have sufficiently proximate repercussions on rights guaranteed by the Convention, even if these repercussions occur outside its jurisdiction'.

and falling back upon more usual default positions, such as the claim that their poor climate mitigation efforts cannot be judged as equivalent to injuries imposed on vulnerable peoples. In this way, law offers a means of mobilizing against multiple agents responsible for the destruction of homelands, food sources, economic livelihoods and traditional ways of life. It also offers a way of conceptualizing territorially dispersed harm agents as co-responsible for the violation of various legally grounded rights obligations.[8]

There is, undoubtedly, a growing trend internationally for citizen alliances to bring increasing pressure to bear on legal authorities to target such transnational sources of harm. In the process, greater consideration is given to the question of how principles of justice might be situated more effectively within a relational framework accounting for multiple agents, causes and effects and acknowledging the importance of geographical, generational and socio-cultural differences between peoples in terms of lived experiences of climate change loss (Skillington, 2019b). Any ambiguities arising in relation to the question of what constitutes an agent of harm (for example, a state, a group of states or corporations) or a 'fair share' of responsibility for transboundary harms can be resolved using methodologies such as the Climate Action Tracker which identifies how much each actor contributes in terms of emissions, both presently and historically, their economic wealth and current per capita emissions. Such methods can easily be used to allocate responsibility in ways consistent with states' 'highest possible ambition' (Paris Agreement, Article 4(3) (United Nations, 2015)) when addressing climate change.

Such an argument was made by six Portuguese youths in September 2020 when they brought a case to the European Court of Human Rights against the EU 27 plus the UK, Norway, Russia, Turkey, Switzerland and Ukraine. The pollution activities of listed states are said to have 'sufficiently proximate repercussions' on the abilities of vulnerable peoples to achieve security, health and self-determination as rights guaranteed by various international conventions. That is, repercussions occurring outside, as much as inside of each of these states' own jurisdiction. Other violations listed include breaches of Article 2 of the European Convention on Human Rights on the right to life said to be diminishing as a consequence of the narrowing margin of appreciation applicable in the area of climate change mitigation. With the support of the Global Legal Action Network (GLAN), the plaintiffs in this case assert a legal defence of their right to be free of the fear of climate

---

[8] In line with Principle 21 of the 1972 Stockholm Declaration, all states, including groups of states, bear multiple liability for climate harms to third parties (see 'Guiding Principles on Shared Responsibility in International Law').

catastrophe (four of the six are survivors of wildfires in the Leiria region of Portugal in 2017).

Basic capabilities to achieve a safe and secure life, physical and emotional well-being, as well as basic material and political control of one's environment (Nussbaum, 2000, pp 72–5) are said to be compromised by the actions of these states. To advance a case for more effective standards of shared responsibility for such harms and greater transparency of decision making, legal campaigners in this instance utilize both international and constitutionalized legal norms. Both serve to strengthen the civic capacities of these citizens to act in relation to transnationally sourced climate harms with a view to achieving democratic redress (Skillington, 2019a, pp 123–30). To strengthen the normative relevance of their claims, current transnational partnerships amongst territorially dispersed agents (that is, the EU or UN community) are highlighted, particularly those who in aspiring to greater co-responsible action must also acknowledge their contributions to the production of transboundary harms and injuries to third parties.

Conclusion

This chapter explores how traditional approaches to the question of responsibility for climate harms come to be subject to critical challenge. More recent years have seen a surge in the number of citizen-led campaigns against various state alliances whose joint pollution activities are said to constitute clear acts of ecological, social and civic injustice. The focus in this instance tends to be on scenarios of 'multiple domination' (Forst, 2020) generated locally, regionally, nationally and internationally by territorially dispersed harm agents and the effects of their actions on the most climate vulnerable (those whose capacities to thrive and adapt to climate change are being steadily undermined). With average temperatures rising steadfastly each consecutive year, Arctic communities are particularly vulnerable to wildfire destruction, deforestation, loss of sea ice, the erosion of settlements and essential habitats (National Oceanic and Atmospheric Administration, 2021). That is, harms generated by the actions of many, transnationally dispersed agents. To counter limited and what are seen increasingly as outdated formulations of legal responsibility for such harms, citizens and political alliances (see, for instance, the Arctic Region Declaration in Preparation for the Global Food Systems Summit (United Nations, 2021)) alike advocate for a civic connections approach where the impacts of pollution practices are assessed more readily in terms of their effects on the rights and circumstances of peoples across borders. The focus thereby shifts from purely local or national impact assessments to multistate level impacts and from merely ecological impacts to those affecting human, cultural, economic, civil and political rights. Responsibility is thereby defined within a web of multiple contexts of subjection to climate harms, as are duties to protect. A principle of proportionality is asserted in the allocation of responsibility for the multiple injuries generated within these contexts and demands are made for a more rigorous regulation of the production and distribution of risks across borders and time (for example, the effects of deep-sea mining of metals, gas and oil in the waters of the North Sea, the acidification of the Arctic's oceans, the pollution impacts of war in the Ukraine, and so on). Arguably, this shift in focus towards the varieties of negative relationalism generated by the pollution practices of several, transnationally dispersed agents renews the institutional relevance of the concept of shared responsibility and creates newly contested grounds for insisting that the principles and terms of internationally agreed treaties, in particular, the Paris Agreement (United Nations, 2015), the international Covenant on Civil and Political Rights (OHCHR, 1966) and the international Covenant on Economic, Social and Cultural Rights (OHCHR, 1966) be implemented more thoroughly in ways that are truly relationally relevant and protective of the interests and needs of vulnerable communities, including those of the Arctic.

Study questions

1. In what ways do climate change related harms intensify the relevance of legal rights norms?
2. Explain how a civic connections approach to climate change differs from a social connections approach.
3. Assess the limitations of a strictly property rights-based approach to natural resource justice.

Acknowledgements

This chapter has received funding from the European Union's Horizon 2020 research and innovation programme under grant agreement No 869327.

The author would like to thank fellow contributors to this volume, Anna Badyina and Darren McCauley for helpful comments in revising this chapter.

References

Arctic Council (2019) 'The Arctic Monitoring and Assessment Programme (AMAP) Arctic Ocean Acidification Assessment 2018: Summary for Policy Makers', [online], Available from: https://www.amap.no/documents/download/3296/inline [Accessed 17 February 2022].

Arctic Resilience Interim Report: Summary for Policy Makers (2013) *Stockholm Environment Institute and Stockholm Resilience Centre*, [online], Available from: https://www.stockholmresilience.org/download/18.416c425f13e06f977b179c/1459560323502/ArcticResilienceInterimReport2013-LowRes.pdf [Accessed 30 August 2021].

Beck, U. (2006) *The Cosmopolitan Vision*, Cambridge: Polity Press.

Carbon Majors (2020) Carbon Majors 2018 Data Set (2020), [online], Available from: https://climate accountability.org/carbonmajors_dataset 2020.html [Accessed 19 July 2021].

Cassotta, S. (2021) 'Ocean acidification in the Arctic in a multi-regulatory, climate justice perspective', *Frontiers Climate*. https://doi.org/10.3389/fclim.2021.713644.

Coello-Camba, A., S. Agusti, J. Holding, J.M. Arrieta, and C.M. Duarte (2014) 'Interactive effect of temperature and CO_2 increase in Arctic phytoplankton', *Frontier Marine Science*. https://doi.org/10.3389/fmars.2014.00049.

Collins English Dictionary (2021) 'Wildfires', [online], Available from: https://www.collinsdictionary.com/dictionary/english/wildfire [Accessed 26 December 2022].

FAO (Food and Agriculture Organisation of the United Nations) (2019) Indigenous Peoples Rome Declaration on the Arctic Region Fisheries and Environment, Available from: https://www.fao.org/uploads/media/FINAL_Rome_Arctic_Declaration_2019_.pdf [Accessed 26 December 2022].

Forst, R. (2017) *Normativity and Power: Analysing Social Orders of Justification*, Oxford: University of Oxford Press.

Forst, R. (2020) 'A critical theory of transnational (in-)justice: realistic in the right way', in T. Brooks (ed.) *The Oxford Handbook of Global Justice*, Oxford: Oxford University Press, pp 452–72.

Galtung, J. (1969) 'Violence, peace and peace research', *Journal of Peace Research*, 6(3): 167–91.

Gardiner, S. (2010) 'A perfect moral storm: climate change, intergenerational ethics, and the problem of moral corruption', in S. Gardiner, S. Caney, D. Jamieson and H. Shue (eds) *Climate Ethics*, Oxford: Oxford University Press, pp 87–98.

Habermas, J. (2008) 'The constitutionalisation of international law and the legitimation problem of a constitution for world society', *Constellations*, 15(4): 444–55.

Katz, C. (2019) 'Why rising acidification poses a special peril for warming Arctic waters', *Yale Environment 360*, [online], Available from: https://e360.yale.edu/features/why-rising-acidification-poses-a-special-peril-for-warming-arctic-waters [Accessed 26 December 2022].

Lèbre, É., M. Stringer, K. Svobodova , J.R. Owen, D. Kemp, C. Cote, et al (2020) 'The social and environmental complexities of extracting energy transition metals', *Nature Communications*, 11(1): 4823. doi:10.1038/s41467-020-18661-9.

NASA (2019) 'Boreal forest fires could release deep soil carbon', [online], Available from: https://climate.nasa.gov/news/2905/boreal-forest-fires-could-release-deep-soil-carbon/ [Accessed 20 January 2022].

National Oceanic and Atmospheric Administration, US Department of Commerce (2021) 'Arctic report card: climate change transforming the Arctic into "dramatically different state"', [online], Available from: https://www.noaa.gov/news-release/arctic-report-card-climate-change-transforming-arctic-into-dramatically-different-state [Accessed 5 August 2022].

Nixon, R. (2013) *Slow Violence and the Environmentalism of the Poor*, Harvard, MA: Harvard University Press.

Nussbaum, M. (2000) *Women and Human Development: The Capabilities Approach*, Cambridge: Cambridge University Press.

OHCHR (1966) International Covenant on Civil and Political Rights. Available from: https://www.ohchr.org/sites/default/files/Documents/ProfessionalInterest/ccpr.pdf [Accessed 26 December 2022].

O'Mahony, P., and T. Skillington (2012) 'Perspectives on cosmopolitanism', *Irish Journal of Sociology*, 20(2): 1–9.

Pettit, P. (1996) 'Freedom as antipower', *Ethics*, 106(3): 576–604.

Rantanen, M., Karpechko, A.Y., Lipponen, A., Nordling, Kalle., Hyvärinen, O., et al. (2022) 'The Arctic has warmed nearly four times faster than the globe since 1979', *Communications Earth & Environment*, 3: 168. https://doi.org/10.1038/s43247-022-00498-3.

Roth, A. (2021) '"Everything is on fire": Siberia hit by unprecedented burning', *The Guardian*, 20 July, [online], Available from: https://www.theguardian.com/world/2021/jul/20/everything-is-on-fire-siberia-hit-by-unprecedented-burning [Accessed 25 July 2021].

Sámi Council (2021) Arctic Region Declaration in Preparation for the Global Food Systems Summit, [online], Available from: https://summitdialogues.org/wp-content/uploads/2021/07/FINAL-ARCTIC-REGION-Declaration-in-preparation-for-the-FSS_29062021.pdf [Accessed 30 July 2021].

Sen, A. (1992) *Inequality Re-examined*, Oxford: Clarendon Press.

Skillington, T. (2017) *Climate Justice and Human Rights*, New York: Palgrave.

Skillington, T. (2019a) *Climate Change and Intergenerational Justice*, Abingdon: Routledge.

Skillington, T. (2019b) 'Changing perspectives on natural resource heritage, human rights and intergenerational justice', *The International Journal of Human Rights*, 23(4): 615–37.

Skillington, T. (2021) 'Natural resource inequities, domination and the rise of youth communicative power', *Distinktion*, 22(1): 23–43.

United Nations (1992a) United Nations Framework Convention on Climate Change, [online], Available from: https://unfccc.int/resource/docs/convkp/conveng.pdf [Accessed 19 July 2021].

United Nations (1992b) United Nations Convention on Biological Diversity, [online], Available from: https://www.cbd.int/doc/legal/cbd-en.pdf. [Accessed 26 December 2022].

United Nations (1997) Kyoto Protocol, [online], Available from: https://unfccc.int/resource/docs/convkp/kpeng.pdf [Accessed 17 February 2022].

United Nations (2015) Paris Agreement, [online], Available from: https://unfccc.int/sites/default/files/english_paris_agreement.pdf [Accessed 17 February 2022].

United Nations (2021) Food Systems Summit, [online], Available from: https://www.un.org/en/food-systems-summit [Accessed 26 December 2022].

Young, I.M. (2011) *Responsibility for Justice,* Oxford: Oxford University Press.

4

A JUST CSR Framework for the Arctic

Darren McCauley

Introduction

Corporate social responsibility (CSR) is the primary mechanism through which private businesses seek to establish their sustainability credentials (Rendtorff, 2019; Saeed et al, 2021). It is supplemented recently with environmental, social and governance investment frameworks (Pedersen et al, 2021). The key concept within both agendas is responsibility. I recognize from the outset that the focus of this chapter is rather narrow in scope due to the word limitations of a chapter. There are other contributions in this book which demonstrate a more critical justice account that moves well beyond the rhetoric of CSR and stakeholders. Chapter 3 is an excellent example of the application of a more critical approach to responsibility (Skillington, 2023). I continue here with a focus on reflecting on and improving the CSR approach of companies through considering five key dimensions of justice: distributional, procedural, recognition, restorative and cosmopolitan.

There is rarely a detailed reflection on what is understood by responsibility within companies reporting activities (Bou-Habib, 2019). A cursory glance at reporting activities reveals that their interpretation is soaked with theoretical and conceptual assumptions around its definition, purpose and elasticity. This chapter is a brief attempt to strengthen the inadequacies of a responsibility focused approach undertaken by private businesses in the Arctic. It argues that responsibility, as understood by companies, is a purposely limiting effort to concentrate on environmental impacts where the Arctic region is concerned. It begins with an introduction to how scholars approach CSR in the Arctic. It will navigate through the different meanings of corporate social responsibility, indicating the ambiguity and divergence of CSR practice in the Russian and Norwegian Arctic. The chapter puts forward a new justice-based framework for CSR with a range of principles that is recommended as a direct consequence of the framework.

Which companies are 'responsible' in the Arctic?

A fruitful initial step is to investigate existing scholarly activities in classifying responsibility. Existing literature on companies in the Arctic is firstly, environmentally focused (Hennchen, 2015; Loe et al, 2017; Overland et al, 2021), and then secondly, livelihoods focused (Wettstein, 2009; Olawuyi, 2016; McQueen, 2019). This is perhaps not surprising considering the environmentally sensitive context in which the Arctic region exists. Responsibility is both short- and long-term, reinforcing the need for businesses to accept the implications of profit-making activities in the region. The environmental impact-oriented approach of responsibility by companies is an especially Arctic observation (Loe et al, 2017). In other words, in regions such as Latin America, we find a more social first approach with regard to Indigenous rights (Ehrnström-Fuentes and Kröger, 2016). This is evident in business and management scholarship in the Arctic, but often in a limited project-based approach. The environmental focus of responsibility by companies is highlighted through the development of an Arctic environmental responsibility index for companies' activities (Overland et al, 2021). Table 4.1 shows what are the most and least responsible companies active in the region.

Overland et al (2021, p 162) define responsibility as 'seeking to avoid harm to the Arctic environment and the culture and livelihoods of Arctic peoples' without any detailed reflection on the implications of such an approach. They go on to detail the following environmental aspects (damage, species, toxins, accidents, clean-up, subcontractors, environmental performance) with legal adherence to minimum rights for Indigenous peoples and company reporting activities. Their definition exposes the limited scope and nature of their responsibility definition. This is partly driven by the limitations of available data sources. But it does reveal little reflection on justice. They find that Norwegian companies lead the way with their approach to responsibility in the Arctic. Equinor, Total and Acker BP are understood to be the most responsible companies in the region. This categorization uncovers the inadequacies (limited scope, definition, data categories) of a responsibility only focus in the Arctic, without admittedly showing the full gamut of negative impacts from corporate entities. There is a need to introduce justice-related arguments into such categorizations to make explicit justice considerations.

Table 4.1: Arctic Environmental Responsibility Index, top and bottom 3 companies

	Top 3			Bottom 3		
Canada	11* Baffinland Iron Mines	12 Chevron	15 MMG Resources	88 True North Gems	91 Northern Cross Energy	108 TMAC Resources
Denmark	31 Capricorn Greenl. Expl.	41 Hudson Resources	54 Bluejay Mining	59 Greenland Resources	66 Nunaoil	84 IronBark Zinc
Finland	8 Anglo American	35 Agnico Eagle Mines	72 Nortec Minerals	98 First Quantum Min.	99 Magnus Minerals	113 Hannukainen Mining
Norway	1 Equinor	2 Total	3 Aker BP	102 Vår Energi	107 Trust Arktikugol	109 The QUARTZ Corp.
Russia	13 Gazprom	19 Kinross Gold	22 Novatek	118 Eriell	119 First Ore Mining	120 Stroygaz Consulting
Sweden	18 Boliden	23 LKAB	101 Beowulf Mining	101 Beowulf Mining	112 EMX Royalty Corp.	115 Sunstone Metals
United States	4 ConocoPhillips	5 BP	6 Exxon Mobil	70 Great Bear Petrol.	94 Brooks Range Petr.	96 Caelus Energy

Note: * Numbers refer to their ranking, 1st being best, 120th worst
Source: Adapted by the author from Overland et al (2021)

Why being responsible is not enough

Responsibility, as defined previously, allows harmful activities with the proviso that remediation can take place. A more extensive definition is found in Sardo (2020, p 73), who concludes that 'agents bear responsibility not in virtue of the individual causal contribution capacity, but because they participate in and benefit from carbon intensive structures, practices and institutions that constitute the global political and economic system'. The first observation from a more extensive definition is why, then, are two of the most active companies in fossil fuel activities in the Norwegian Arctic defined by Overland et al (2021) as the most responsible? Being responsible for clean-ups, or supply chains, merely upholds structural injustice in the

Arctic and, moreover, globally. The second observation that follows is that responsibility, as narrowly defined, is incapable of addressing the injustices enacted by fossil fuel companies, never mind helping global societies to consider a post-fossil fuel world.

Responsibility is in doubt as a useful concept for guiding companies' behaviours in the Arctic. Current scholarship in practice suggests that it limits the moral and ethical dimensions of introspection (Szczepankiewicz and Mućko, 2016; Rendtorff, 2019). A wider framework of justice-based thinking is better placed to achieve more sustainable actions from private entities in the Arctic. Before outlining a JUST CSR framework, which is basically a new framework of justice-based principles to be applied to CSR, I take a brief look at how the broader concept of CSR is implemented by energy companies in the Arctic, with a focus on Russian firms. Russian companies are selected as least well-performing as identified in Table 4.1. The role of stakeholders and shareholders is raised in addition to the divergence of existing CSR practices in the energy sector. I have argued that responsibility is ineffective by itself for achieving sustainable behaviours. I develop this thesis in the next section with further inadequacies raised in terms of the limited corporate view of whose interest should be promoted and the divergent practices undertaken by energy companies in the name of responsibility.

Energy companies and CSR in the Arctic

There is a well-established literature that argues that CSR helps companies to develop their internal sustainability, diversity and inclusion policies (Shahbaz et al, 2020; Karaman et al, 2021). I focus on the second dynamic of CSR and focus on the external impacts of corporate behaviour (Hennchen, 2015; Olawuyi, 2016). It is here where we find most disputes in existing literature, as well as in practice, as I will briefly outline.

The inadequacies of a 'share-/stake-holder first' approach to society

The primary question for assessing the approach of energy companies towards the issue of responsibility within the CSR framework is: who do private firms view as those affected by their actions? The external impacts of energy companies are determined by the ways in which such actors define the limitations of those who are affected (for further discussion on externalities in this field, see Price, 2007, and Bellanger et al, 2021). In CSR literature, this is referred to as stakeholder theory (Doh and Guay, 2006; Banerjee and Bonnefous, 2011). This approach is in fact an extension of the original framework for understanding corporate behaviour through shareholder focus behaviours (George, 2019). This initial idea is that companies should prioritize the interests only of their shareholders. Scholars criticized this approach as leading to environmental damage and the wide sweeping ignorance of communities impacted upon by irresponsible corporate behaviour (Rönnegard and Smith, 2013). Stakeholder theory is therefore positioned as the CSR answer. This means that companies should consider not only the shareholders but classify and report upon impacts to a wide range of stakeholders from local communities to overlooked Indigenous groups.

The categorization and classification of stakeholders limit the concept of responsibility. It has developed into a means for classifying who is worthy of corporate reflection, and who is not. This has resulted in some scholars arguing that the environment itself should be perceived as a stakeholder (Cotton and Mahroos-Alsaiari, 2015). However, this avoids the necessary reflection on the ethical and moral standards inherent in a company's approach towards its activities. The stakeholder mechanism encourages a tick box mentality by incentivizing companies to undertake a stakeholder analysis before continuing to pursue their original project. This is, of course, better than nothing. But if our objective is better than nothing, the Arctic, the climate, and our future are surely under threat.

The ambiguity and divergence of CSR practices in the Arctic

I briefly provide examples of how energy companies active in the Arctic interpret CSR. As mentioned earlier, Russian companies are the least well-performing (see Table 4.1) in relative terms to Norwegian companies. Gazprom is selected as the best performing, and Lukoil is a moderately performing company (the lowest performing has insufficient documents available). There is insufficient space in this chapter to elaborate in depth, so this is not presenting systematic empirically researched data. It is, rather, presenting examples of two prominent Russian companies and their different strategic approach to implementing CSR. I completed a frequency of terms analysis on the main CSR documents of each company. In the first case, Lukoil sees CSR as a frame through which to consider driving efficiency and environmental protection through their external activities. Gazprom follows an understanding that prioritizes technological development and health impacts. In short, CSR is loosely applied – leading to divergence with little reflection on the values that underpin each strategy.

Lukoil strategy on CSR – efficiency and environmental best practice

Lukoil is actively engaged in developing Arctic projects in collaboration with other companies on joint ventures. I do not seek to investigate such joint ventures here, but rather to expose the way in which the company views CSR through a frequency of terms analysis. It is apparent from existing documents that energy efficiency and environmental best practice are the two key objectives for this organization in implementing CSR policies. A brief examination of a CSR strategy in 2021 reveals that 'we see our main task as efficient reinvestment of capital into production expansion' (Lukoil Group, 2021, p 32). The report goes on to establish a range of environmental best practice procedures which includes 'making significant investments in industrial safety and environmental projects and demonstrat[ing] excellent results in reducing the frequency of accidents and environmental stress' (Lukoil Group, 2021, p 46). If we look beyond its formal CSR strategy, we can see this approach also in the health safety and environmental policy. It states that 'while being aware of the social responsibility, the company intends to contribute to long-term economic growth through preserving favourable environmental conditions and ensuring the efficient utilisation of natural resources' (Lukoil, 2021).

The purpose here is not to undertake a systematic investigation of existing reports, but rather to outline the key principles at play for this company concerning CSR by exposing which terms come up the most in their CSR documents. It is evident from this cursory investigation that efficiency and environmental best practices are the core objectives. There is little evidence in these documents of consideration to technology or health matters.

Gazprom Neft strategy on CSR – modernization, technology and health

In contrast, the Gazprom approach to CSR differs greatly from that of Lukoil. We do not see the same level of reflection upon efficiency and environmental best practice through the frequency of terms analysis. The focus of their CSR activities is predominantly on modernization through the developments and implementation of new technology. A secondary focal point for this company is around developing health policies. The closest document to a formal CSR policy was published in 2020, referred to as its sustainability report. The organization states throughout the document that it 'pays close attention to developing the best available technologies' (Gazprom Neft, 2020, p 16). It develops a wide range of technology focused activities on what is referred to as 'corporate social responsibility priorities'. Moreover, their policy is clearly connected to nationalistic and conservative principles, often referring to 'technology for bringing energy security to Russian communities, and furthering traditional value systems' (Gazprom Neft, 2020, p 32). Their secondary area of health appears on several occasions with regards to maintaining healthy environments for Russian communities as well as workers in industry.

The two brief case studies based on a basic frequency of terms analysis reveal different approaches towards enacting what the company views as responsibility. There is clear evidence of a stakeholder, rather than shareholder approach, in such documents, where allusions are used to a wide range of affected entities such as communities or even the physical environment. It reinforces further the voluntary nature of CSR. Responsibility lies with the company to define its core principles. It effectively leads to further profit-driven actions, rather than any awareness of its environmental and social impacts. For example, the framing of developing technology or energy efficiency in terms of environmental impacts promotes further revenue streams for companies. A new approach is consequently needed.

A JUST framework for CSR in the Arctic

Global demand for Arctic resources is set to increase post-pandemic, and, with it, companies are expected to contribute to the sustainability of the environment as well as society. Throughout these supply chains, we expect healthy workplaces, fair payment schemes, tailored services and consumer protection (Nurunnabi et al, 2020). We also expect a high level of social responsibility beyond a company's immediate supply chain concerns (Shahbaz et al, 2020), demonstrated further in global shipping research (Kitada and Ölçer, 2015). Hamilton (2011) outlined four key areas of concern: *translation*, where social and environmental problems have been notoriously difficult to translate in terms of support for CSR initiatives; *dilution*, when flexible supply chains avoid increasing levels of regulations or standards; *access*, insofar as some producers throughout the supply chain cannot afford to maintain increasing standards; and *embedding*, where the local institutional frameworks in the Arctic can negatively impact on CSR initiatives. This all points towards the inadequacies of CSR to develop sustainable solutions across supply chains in the Arctic, and highlights the need for a new framework in this area.

CSR is by itself an insufficient framework for ensuring that global supply chains are sustainable for the Arctic. New ideas are urgently needed. Existing work has demonstrated that supply chains move around different constituencies, with the fact that regulations do not consistently apply. The Arctic region epitomizes this issue, but it does not stand alone. Healy et al (2019) demonstrate how one such supply chain from Salem, Massachusetts, connects in such a way with an open pit coal mine in Colombia. Carpenter and Wagner (2019) reveal in a different way that the oil refinery industry across the US has similar impacts on different communities which are disproportionally suffering from economic inequality. The reporting practices at the heart of CSR are simply inadequate, highlighted also in research on Polish energy and mining companies for example (Szczepankiewicz and Mućko, 2016). These impacts of CSR are not limited to fossil fuels. Heffron and McCauley (2014) detail how global wind energy systems have similar impacts and have remained outside the scope of CSR, a finding supported by other research in this area (Mezher et al, 2010). This chapter argues in line with other scholars (Manteaw, 2008; Hamilton, 2011; Weber, 2018;

Rendtorff, 2019) that a new responsible approach is needed that goes beyond CSR for the Arctic; it differs, however, in calling for such an approach to place justice at the centre.

Justice frameworks hold the key to ameliorating the inadequacies of CSR. To begin, social responsibility must be replaced and fortified through including social justice (Shaw, 2016). Responsibility is a term that is too often focused on process rather than outcome (Wettstein, 2009). Scholars (Mutch and Aitken, 2009; Bakhtina and Goudriaan, 2011) have argued that a company's responsibility can be fulfilled through new ways of completing social and environmental checklists or adhering to basic regulatory frameworks on the ground. The Arctic region is an archetypal example of this approach. This has the effect of driving minimal action to maintain reputational value throughout supply chains. Social justice encompasses a wider demand to adhere to both processes and outcomes (Newell and Frynas, 2007). Manteaw (2008) details how comprehensive assessments of justice theory can result in more complex geographically sensitive and effective insights into corporate behaviour. Social justice is often reduced to a Kantian or Rawlsian critique which suffers from the same process-driven dominated view as responsibility (Amalric et al, 2004; Newell and Frynas, 2007). This chapter interjects in this debate to put forward and assess how justice frameworks could help to ensure a wider view taken by companies. It allows us to move beyond volunteering, beyond CSR and beyond even narrow applications of human rights. It challenges us to reflect on how global companies can help drive a sustainable climate transition for the Arctic (Arnold, 2013).

What is the JUST CSR framework?

The JUST CSR framework is an alternative approach to existing company focused conceptualizations. The JUST (Justice, Universal, Space and Time) framework has previously been applied to legal frameworks (Heffron and McCauley, 2018) and more recently to critique the emergence of energy democracy as a new agenda (Droubi et al, 2022). The novelty here is the explicit application to CSR and private companies. This application is new and offers much potential for new scholarly and practice-oriented responses. Existing research has demonstrated its practical applicability to the implementation of a more just approach towards critical minerals (Heffron, 2020). The framework builds upon conceptual work outlined in McCauley and Heffron (2018). It set out the basis of the JUST framework in Droubi et al (2022) as an opportunity to bring together climate, environmental and energy justice scholarship. Combined, the resulting JUST CSR framework allows the researcher and practitioner to consider several issues from the external practicalities of supply chains to more critical reflection on internal activities such as diversity and inclusion policies. The JUST framework has evolved to incorporate five distinct areas of justice:

- *Distributional justice.* From a company perspective, this dimension of justice sheds light on where the benefits and ills of their activities lie. This application encourages the company to consider precisely where impacts are most felt.
- *Procedural justice.* Due process and adherence to legal based rights in representation and decision making are necessary components of a company's activity. Concretely, this leads the company to consider, for example, its impact assessments, from environmental to social and economic, in more depth.
- *Recognition justice.* This is a post-distributional form of justice where focus from the company is placed on which different parts of society are impacted upon. This is most useful for considering the recognition of Indigenous communities' rights.
- *Cosmopolitan justice.* This is built upon the foundations of global justice that consider the need to establish a basic level of rights conferred upon all citizens of the world. It means that a company should recognize the role of rights protection beyond individual national legal constituencies, moving beyond national borders.
- *Restorative justice.* A company must reflect on how it not only undertakes its practices in a fair and equitable manner, but also how it will systematically set about restoring the negative implications of its past activities. From an energy perspective, this has a wide range of applications from previous extractive activities to the moral responsibility of decommissioning.

The JUST CSR framework is complemented by the application of four additional dimensions to those mentioned earlier, as detailed in the JUST framework set out in Droubi et al (2022). The first is referred to as *Justice*. The justice aspect of the framework refers to the 'particular' (LaBelle, 2020) forms of justice – distributional, procedural and restorative justice. The *Universal*, as defined by LaBelle (2020), aspect of the framework refers to global rights such as cosmopolitan and recognition justice. These are more holistic forms of justice. The third aspect of the framework involves *Space*. This is an explicit consideration of the level at which a benefit or ill is experienced, whether it be local, national or global. The last component of the framework is referred to as *Time*. A company from this perspective needs to think about explicitly the past and future in a way that analyses its current approach towards the transition and subsequent policies generated for ensuring fairness at each point in time, 1990, 2020, 2050, and so on.

What are the implications of a JUST framework on a company's activities in the Arctic?

The JUST CSR approach means that companies can follow a set of clear universal principles to guide their CSR activities in the Arctic. The divergences in practices, coupled with debatable applications and practice, mean that a justice-based approach is needed. As a starting point, I provide five clear principles which should guide companies reporting activities for Arctic based projects. These principles are not designed to be all encompassing or to replace CSR. They do not, for example, demand radical actions like halting all extractive activities. The principles are best viewed as high level principles that should guide the re-organization of existing CSR policies and practices. The development of such a standardized approach would drive more sustainable behaviours whilst allowing the company to reflect more meaningfully on its activities at project locations, both in and beyond the Arctic.

- *Principle 1 – Distribute the benefits and minimize burdens of a project locally.* Each company should ensure that there is sufficient data and analysis, before activities are undertaken, on where the benefits and burdens of any given project will fall. This reinforces the need to implement a common approach towards environmental, social and economic impact assessments with the stated objective of identifying how benefits can be shared and burdens minimized.
- *Principle 2 – Comply with legal regulations and engage with affected communities.* Companies need to show a more systematic awareness of legal and regulatory demands on all three sectors of sustainability, namely environment, society and the economy. This awareness must be explicitly connected with both formal and informal regulatory commitments, such as recognizing and respecting Indigenous legal systems and cultures.
- *Principle 3 – Prioritize Indigenous communities' rights.* Companies should clearly state how Indigenous communities will benefit from any given project. There should be an adoption of an 'Indigenous communities first' approach where proposed activities are measured against the direct and indirect benefits for such communities. This should be developed in coordination with local representatives.
- *Principle 4 – Engage in making sure any given project has a net positive global impact.* Unlike existing environmental impact assessment obligations, companies should consider their broader global impact. This means that a clear indication of its wider impact beyond the Arctic must be noted, for example detailing a project's impact across the supply chain in other jurisdictions beyond the Arctic.

- *Principle 5 – Undertake a formal obligation to be accountable for future negative impacts.* Each company should consider the development of formal structures and processes of obligation to restoring past, present and future negative damage. This includes environmental, social and economic negative impacts on affected communities.

Conclusion

Being responsible is not enough. The Arctic Environmental Responsibility Index outlined in the Introduction is a warning. Undertaking responsible processes appear to be the end game. But the Arctic needs positive outcomes. Responsibility for assessing or voluntarily reporting on environmental damage, species impacts, toxins release, accidents, clean-up, subcontractors and overall environmental performance is only a beginning. The inclusion of justice scholarship in the activities undertaken by private companies expands beyond the concept of responsibility. Companies are in this way encouraged to think about both fairness and equity in process and outcome. They should better account for where they implement their projects and ensure that their consequent impacts on inequalities are positive. Adherence to both international legal norms, and informal Indigenous legal systems, must be reached. Taking a responsible approach is to consider Indigenous rights. But being just in this regard means pursuing an Indigenous rights first guarantee. Responsibility cannot end at one village, town, region or even nation. JUST CSR in this way requires global-level consideration and positive outcomes. Lastly, the future is not ours to see. It is ours to protect. Intergenerational restorative justice needs to be a key principle of corporate behaviour in the Arctic. Corporate social responsibility has the chance to reform. Justice is that chance.

Study questions

1. What is the JUST framework and how does it interact with CSR?
2. Can you develop different principles to the five highlighted previously from the JUST CSR framework?
3. What factors like those developed in the 'Arctic Environmental Responsibility Index' would you include in a 'JUST CSR Index'?
4. Select a non-energy company that is active in the Arctic. How would you define their approach to responsibility? How could it be improved by applying the JUST CSR framework?

Acknowledgements

This chapter has received funding from the European Union's Horizon 2020 research and innovation programme under grant agreement No 869327.

References

Amalric, F., D. Kennedy-Glas, S. Reddy, M. O'Sullivan, and J. Trevino (2004) 'Can CSR make a contribution to international solidarity and the quest for social justice in the South?', *Development*, 47(3): 38–46.

Arnold, D.G. (2013) 'Global justice and international business', *Business Ethics Quarterly*, 23(1): 125–43.

Bakhtina, K., and J.W. Goudriaan (2011) 'CSR reporting in multinational energy companies', *Transfer: European Review of Labour and Research*, 17(1): 95–9.

Banerjee, S.B., and A.M. Bonnefous (2011) 'Stakeholder management and sustainability strategies in the French nuclear industry', *Business Strategy and the Environment*, 20(2): 124–40. doi:10.1002/bse.681.

Bellanger, M., R. Fonner, D. Holland, G. Libecap, D. Lipton, and P. Scemama (2021) 'Cross-sectoral externalities related to natural resources and ecosystem services', *Ecological Economics*, 184: 106990. doi:10.1016/j.ecolecon.2021.106990.

Bou-Habib, P. (2019) 'Climate justice and historical responsibility', *The Journal of Politics*, 81(4): 1298–310.

Carpenter, A., and M. Wagner (2019) 'Environmental justice in the oil refinery industry: a panel analysis across United States counties', *Ecological Economics*, 159: 101–9.

Cotton, M.D., and A.A. Mahroos-Alsaiari (2015) 'Key actor perspectives on stakeholder engagement in Omani Environmental Impact Assessment: an application of Q-Methodology', *Journal of Environmental Planning & Management*, 58(1): 91–112. doi:10.1080/09640568.2013.847822.

Doh, J.P., and T.R. Guay (2006) 'Corporate social responsibility, public policy, and NGO activism in Europe and the United States: an institutional-stakeholder perspective', *Journal of Management Studies*, 43: 47–73.

Droubi, S., R.J. Heffron, and D. McCauley (2022) 'A critical review of energy democracy: a failure to deliver justice?', *Energy Research & Social Science*, 86: 102444. doi:10.1016/j.erss.2021.102444.

Ehrnström-Fuentes, M., and M. Kröger (2016) 'In the shadows of social licence to operate: untold investment grievances in Latin America', *Journal of Cleaner Production*, 141: 346–58. doi:10.1016/j.jclepro.2016.09.112.

Gazprom Neft (2020) *Sustainable Development Report 2020*, Gazprom Neft, [online], Available from: https://www.gazprom-neft.com/annual-reports/2020/csr_en_annual-report_pages_gazprom-neft_2020.pdf [Accessed 23 November 2021].

George, E. (2019) 'Shareholder activism and stakeholder engagement strategies: promoting environmental justice, human rights, and sustainable development goals', *Wisconsin International Law Journal*, 36(2): 298–365.

Hamilton, T. (2011) 'Putting corporate responsibility in its place', *Geography Compass*, 5(10): 710–22.

Healy, N., J.C. Stephens, and S.A. Malin (2019) 'Embodied energy injustices: unveiling and politicizing the transboundary harms of fossil fuel extractivism and fossil fuel supply chains', *Energy Research & Social Science*, 48: 219–34.

Heffron, R.J. (2020) 'The role of justice in developing critical minerals', *Extractive Industries and Society*, 7(3): 855–63. doi:10.1016/j.exis.2020.06.018.

Heffron, R.J., and D. McCauley (2014) 'Achieving sustainable supply chains through energy justice', *Applied Energy*, 123: 435–7.

Heffron, R.J., and D. McCauley (2018) 'What is the "Just Transition"?', *Geoforum*, 88: 74–7. doi:10.1016/j.geoforum.2017.11.016.

Hennchen, E. (2015) 'Royal Dutch Shell in Nigeria: where do responsibilities end?', *Journal of Business Ethics*, 129(1): 1–25.

Karaman, A.S., N. Orazalin, A. Uyar, and M. Shahbaz (2021) 'CSR achievement, reporting, and assurance in the energy sector: does economic development matter?', *Energy Policy*, 149: 112007. doi:10.1016/j.enpol.2020.112007.

Kitada, M., and A. Ölçer (2015) 'Managing people and technology: the challenges in CSR and energy efficient shipping', *Research in Transportation Business & Management*, 17: 36–40.

LaBelle, M.C. (2020) *Energy Cultures: Technology, Justice, and Geopolitics in Eastern Europe*, London: Edward Elgar Publishing.

Loe, J., I. Kelman, D. Fjærtoft, and N. Poussenkova (2017) 'Arctic petroleum: local CSR perceptions in the Nenets region of Russia', *Social Responsibility Journal*, 13(2): 307. doi:10.1108/SRJ-10-2015-0150.

Lukoil (2021) *Health, Safety and Environment Policy of Lukoil Group in the 21st Century*, [online], Available from: http://www.lukoil.com/en/HSEPolicy [Accessed 1 April 2021].

Lukoil Group (2021) *Key Changes and Results – Climate Change – Lukoil Group 2019 Sustainability Report*, [online], Available from: https://csr2019.lukoil.com/climate-change/major-changes-and-results [Accessed 1 April 2021].

Manteaw, B. (2008) 'From tokenism to social justice: rethinking the bottom line for sustainable community development', *Community Development Journal*, 43(4): 428–43.

McCauley, D., and R. Heffron (2018) 'Just transition: integrating climate, energy and environmental justice', *Energy Policy*, 119: 1–7. doi:10.1016/j.enpol.2018.04.014.

McQueen, D. (2019) 'Frack off: climate change, CSR, citizen activism and the shaping of national energy policy', in F. Farache, G. Grigore, A. Stancu and D. McQueen (eds) *Responsible People*, Cham: Springer, pp 175–98.

Mezher, T., S. Tabbara, and N. Al-Hosany (2010) 'An overview of CSR in the renewable energy sector: examples from the Masdar Initiative in Abu Dhabi', *Management of Environmental Quality: An International Journal*, 21(6): 744–60.

Mutch, N., and R. Aitken (2009) 'Being fair and being seen to be fair: corporate reputation and CSR partnerships', *Australasian Marketing Journal (AMJ)*, 17(2): 92–8.

Newell, P., and J.G. Frynas (2007) 'Beyond CSR? Business, poverty and social justice: an introduction', *Third World Quarterly*, 28(4): 669–81.

Nurunnabi, M., J. Esquer, N. Munguia, D. Zepeda, R. Perez, and L. Velazquez (2020) 'Reaching the sustainable development goals 2030: energy efficiency as an approach to corporate social responsibility (CSR)', *GeoJournal*, 85(2): 363–74.

Olawuyi, D.S. (2016) 'Climate justice and corporate responsibility: taking human rights seriously in climate actions and projects', *Journal of Energy & Natural Resources Law*, 34(1): 27–44.

Overland, I., A. Bourmistov, B. Dale, S. Irbacher-Fox, J. Juraev, E. Podgaiskii, and F. Stammler (2021) 'The Arctic Environmental Responsibility Index: a method to rank heterogenous extractive industry companies for governance purposes', *Business Strategy and the Environment*, 30(4):1623–43. doi:10.1002/bse.2698.

Pedersen, L.H., S. Fitzgibbons, and L. Pomorski (2021) 'Responsible investing: the ESG-efficient frontier', *Journal of Financial Economics*, 142(2): 572–97.

Price, C. (2007) 'Sustainable forest management, pecuniary externalities and invisible stakeholders', *Forest Policy and Economics*, 9(7): 751–62. doi:10.1016/j.forpol.2006.03.007.

Rendtorff, J.D. (2019) 'The principle of responsibility: rethinking CSR as SDG management', in *Philosophy of Management and Sustainability: Rethinking Business Ethics and Social Responsibility in Sustainable Development*. Bingley: Emerald, pp 205–20.

Rönnegard, D., and N.C. Smith (2013) 'Shareholders vs. Stakeholders: How Liberal and Libertarian Political Philosophy Frames the Basic Debate in Business Ethics', *Business & Professional Ethics Journal*, 32(3/4): 183–220.

Saeed, A., U. Noreen, A. Azam, and M. Tahir (2021) 'Does CSR governance improve social sustainability and reduce the carbon footprint: international evidence from the energy sector', *Sustainability*, 13(7): 3596. doi:10.3390/su13073596.

Sardo, M.C. (2020) 'Responsibility for climate justice: political not moral', *European Journal of Political Theory*, 22(1): 26–50.

Shahbaz, M., A. Karaman, M. Kilic, and A. Uyar (2020) 'Board attributes, CSR engagement, and corporate performance: what is the nexus in the energy sector?', *Energy Policy*, 143: 11–29.

Shaw, C. (2016) 'The role of rights, risks and responsibilities in the climate justice debate', *International Journal of Climate Change Strategies & Management*, 8(4): 505–17.

Skillington, T. (2023) 'Protecting the Arctic's wild lands: a relational view of responsibility', in C. Wood-Donnelly and J. Ohlsson (eds) *Arctic Justice: Environment, Society and Governance*, Bristol: Bristol University Press.

Szczepankiewicz, E.I., and P. Mućko (2016) 'CSR reporting practices of Polish energy and mining companies', *Sustainability*, 8(2): 12–29.

Weber, G. (2018) 'Present approaches and tendencies in sustainable energy strategies in relation to CSR', in *Sustainability and Energy Management*, Cham: Springer, pp 83–100.

Wettstein, F. (2009) 'Beyond voluntariness, beyond CSR: making a case for human rights and justice', *Business and Society Review*, 114(1): 125–52.

5

Collective Capabilities and Stranded Assets: Clearing the Path for the Energy Transition in the Arctic

Roman Sidortsov and Anna Badyina

Introduction

Several years after the signing of the Paris Agreement, oil and gas production continues at a fast pace despite a near global recognition of the ongoing energy transition away from fossil fuels (IEA, 2021a, 2021b). The increasing demand for oil and gas caused by the post-pandemic lockdown economic recovery threatens the momentum gained by the energy transition. The momentum is further threatened by Russia's military aggression in Ukraine and the corresponding bans and restrictions of many oil and gas importing nations on Russian hydrocarbons. While the rapidly increased prices and supply shortages only strengthen the energy transition case, the interdependencies and inertia of the current energy system dictate that the current energy crisis is resolved largely by rearranging the supply options and increasing oil and gas production elsewhere, including the Arctic.

In the early 2000s, the Arctic became an increasingly coveted region for finding and developing new hydrocarbon resources, prompted by the historic production and yet to be confirmed estimates (Sidortsov, 2016; Wood-Donnelly and Bartels, 2022). Several factors make oil and gas development in the region an uncertain, if not reckless, enterprise. These include: (1) legal and regulatory risks due to existing and prospective carbon controls; (2) rapidly changing Arctic landscapes due to climate change, which subjects oil and gas infrastructure to significant physical impacts; (3) high capital costs needed to develop remote and challenging fields; (4) oil and gas price volatility; and (5) the uncertain geopolitical position and problematic political, economic and social situation in the Arctic's largest producer of hydrocarbons, Russia (Wood-Donnelly and Bartels, 2022). Some of these risks, such as shut-in oil wells due to thawed permafrost, have already impacted the ongoing oil and gas activities, resulting in damage to ecosystems (Sidortsov and Gavrilina, 2018). With accelerating physical and policy changes, expansive and costly oil and gas production, transportation and processing facilities are on track to become stranded assets impacting the socio-economic fabric of Arctic communities and fragile Arctic ecosystems (Sidortsov, 2012).

The overarching objective of this chapter is two-fold: (1) to make a case for the inclusion of collective capabilities (CCs) as a key element of analytical and conceptual energy justice frameworks; and (2) to underscore the value of CCs for conceptualizing and assessing the impact of fossil fuel-centric, capital-intensive and long-term economic projects and programmatic activities on the energy transition in the Arctic and other historic oil and gas producing regions.

Background

Oil and gas development is emblematic of the historic approaches to industrial-scale resource extraction in the Arctic region. Projects are often part of larger programmatic initiatives coordinated and/or otherwise supported by governments through subsidies, infrastructure development and various information and persuasion campaigns (Fjaertoft and Lunden, 2015). In addition to the ostensibly positive overall economic effect, sovereign entities and private actors enter agreements that set forth the latter contributions to regional and local development. These public–private collaborations happen at macro (national), meso (regional) and micro (local) levels, often simultaneously. The initiatives create interdependences between subnational and national governments and the industry aligning their interests and committing their financial and organizational resources toward the same goal.

The alignment is often portrayed and, as a result, perceived as a mutually beneficial process of value co-creation. After all, the Arctic region often lacks capacities, expertise and resources due to its remoteness vis-à-vis economic and scientific centres. There are plentiful examples of collaborative practices that result in positive and negative outcomes for sovereign entities (see Chapter 11 on urban transformation in Hammerfest). Practices that lead to positive results are often associated with the high level of a company's *investment* in the relevant community, high social impact standards, extensive mapping, engaging with a range of social actors, and close collaboration between the central and local governments.

There are still many unknowns that get in the way of making the oil and gas industry's regional and local participation in public life an overall positive practice. However, to a large extent the success of such private–public collaborations depends on the availability and quality of what can be defined as *collective politics* that involves deliberative modes of democratic engagement. The existing politics of value creation often lacks the collective foundation and, as a result, is limited in space and time, focusing largely on individual projects and short- to medium-term effects. This conceals that *value* is not just a thing (an object) but above all a social (and conflict-driven) category. As such, the collective politics of value creation can only be properly established and realized through extensive, multi-level and proactive public discourse and decision making. It must work with the processes that highlight the existing and future socio-political realities in the Arctic. Additionally, it requires the public sector (the government) to have appropriate arrangements and capacities. Finally, and perhaps most importantly, it must acknowledge and operationalize the structural limits of the private–public convergence.

These limits are premised on the notion that although the private–public sector convergence might seem as *operationally* beneficial for both sides, their *strategic* interests and objectives are likely to be different (Vizhina et al, 2013). Extractive projects come with inherent expiration dates whereas societies do not. Companies, especially large oil and gas enterprises, can move around the world seeking another profitable venture, but nations, provinces and municipalities cannot do the same. The latter must make do largely with the human, natural and financial capital that they have, while adapting to and mitigating the changes within their borders.

From practical wisdom to individual and collective capabilities

To paraphrase Amartya Sen, the aim of the human, natural and financial capital is the access and opportunity of these societies 'to do things that they have reason to value', or to have collective capabilities (CC) (Evans, 2002, p 55). The capabilities approach (CA) upon which CCs are premised has predominately been used as a method for evaluating human well-being. Both Sen (1999), Martha Nussbaum (2002), and Ingrid Robeyns (2005) develop their takes on the CA through the lens of an individual. Thus, having the capability to function means that there are 'various combinations of functionings (beings and doings) that the person can achieve reflecting the person's freedom to lead one type of life or another' (Sen, 1992, p 40). Thus, people's 'capability to function' in the face of a changing economic, social and environmental conditions is contingent on the 'effective opportunities ... to undertake the actions and activities that they want to engage in and be whom they want to be' (Robeyns, 2005, p 95). This might include moving away from an extraction-based economy but continuing to live (and flourish) in the same place. Because under the conventional CA the 'effective opportunities' are attached to an individual, the aforementioned human, natural and financial capital of sovereign entities would be an aggregate of individual capabilities.

The CA's contextualization of what is good, fair and just in individual (human) and collective (social) circumstances is hardly a modern invention. The roots of CA can be traced back to ancient Greek thinking, primarily intellectuals such as Aristotle. According to Peter Massingham, Aristotle sees the grounding of general prescriptive principles of right and wrong in terms of phronesis or 'practical wisdom', which is to him 'the highest intellectual virtue' or 'the master virtue' premised on 'the complicated interactions between general (theory) and practical (judgement)' (Massingham, 2019, p 2). In other words, one also needs the right means toward good ends, which depend on their knowledge or perception of the particular circumstances in which those prescriptive principles are implemented; if one's knowledge or perception is wrong, one can fail to achieve the good life as defined by such principles. It is in this way that justice can, and often does, differ from one case to another, making it difficult if not impossible to define. For Aristotle, *practical wisdom* cannot be explained like general knowledge, because it is too particular. General justice principles will rarely be directly applicable to real life situations. Instead, our moral knowledge can only be obtained through *experience*. With practical wisdom, we intuitively grasp the particular aspects of the situation. It does not make this exercise subjective because there is a truth of the matter to be known in the end.

General justice principles undoubtedly can and should contribute to setting the ends. Whether it is the prioritization of the least fortunate, universal treatment regardless of one's origin, or maximization of pleasure or minimization of suffering, these principles are indeed the contours of individual or societal flourishing. However, these ends are not sufficient to perform a just action, because they do not define what is just in a particular situation. Following Aristotle, humans require *practical wisdom* to establish the constitutive means to our moral ends. It is necessary to distinguish between acting according to a virtue and performing a completely virtuous action (Massingham, 2019; Nussbaum, 1995). The latter is based on properly grasping what is just in a particular situation and what is required to achieve justice.

Human *capabilities* are, thus, a closely related concept since one's ability to define and achieve the valuable outcomes in their life depends on the situation/conditions in which they are living. Furthermore, advancing capabilities needs to engage with human circumstances, which are seen to involve both *internal* (individual) and *external* (contextual, societal) change factors (Sen, 1985). Capabilities scholars have been concerned with defining a variety of *beings* and *doings* (or functionings) and the real possibilities to achieve those beings and doings (or capabilities) which can be used to make interpersonal comparisons of outcomes and abilities to pursue those outcomes.

However, while defining a set of functioning and capabilities to achieve the good life and justice is essential, it may not be sufficient when it comes to their actual application in a particular scenario. The good life and the possibility to achieve it will differ from one case to another since it is contingent on a range of factors, forces, preferences and social relations at play in the case in question. It will therefore require further *empirical substantiation* or what can be defined as *learning through experience* by examining the surrounding situation (its elements, functions, structures and linkages).

The empirical substantiation is a difficult exercise to implement because of many impact factors that cannot be controlled and managed with the capacity of a single individual. It requires interacting and working with others. Therefore, defining a set of functionings and capabilities completely and achieving justice fully (that is, what it is in a particular situation of people) will depend on the collective moral attitude of a group and the being-with-others and acknowledging other's perceptions and experiences.

Highlighting the ancient philosophical ideas and the roots of the capabilities approach in combination with the approach's current challenges reveals the importance of shifting the focus from an aggregate of individual functionings and capabilities to the collective processes defining and perfecting them. The process is about the continual sharing, debating and learning through each other's *experience* to improve one's understanding of the situation – what is valued and how best to act on the valued outcomes. It is about a type of *democratic collective* that allows people to engage in self-criticism, deliberation and struggle in relation to their common concerns and experiences. Creating capabilities should also recognize the good life as embedded into and constitutive of the world of interrelated and interacting components and processes. It becomes an episode within the story of society that both pre-exists and endures. Capabilities or justice cannot be created and guaranteed apart from the wider societal considerations.

The previously stated point on the importance of collective deliberation connects well with an old sociological tradition that recognizes human existence as imperfectly programmed. Human life is thus an open system which is shaped through human practice. As pragmatist philosophers also highlight, *practice* functions as a medium of truth or the 'quest for cognitive certainty' while serving as an 'inexhaustible source of inspiration for knowledge' (Dewey, 1929, p 21). Practice and knowledge provide 'means of making goods – excellences of all kinds – secure in experienced existence' (Dewey, 1929, p 21). In this sense, practice also essentially means the possibility of having encounters with others. It is from this perspective that we see that humans can know *fully* the situation in which they are living and design and achieve enduring good outcomes only through their mutual experience and work.

The individual and collective intertwine into one, complementary and inseparable, or as Peter Evans (2002, p 55) argues, the access and opportunity of individuals 'to do things that they have a reason to value' depend on *collective* decisions about the distribution of impacts, risks and benefits among different members of a collective body. These collective decisions are, therefore, instrumental to acquiring and formulating individual capabilities. To sum up, 'gaining the freedom to do the things that we have reason to value' is a function of individual and collective capabilities with the latter being not an aggregate or sum of different capabilities, but a combination of mutually constitutive elements. CCs are shared freedoms through interactions and connections acting as a procedural and substantive glue that can bind individual actions into a collective one whilst providing and allocating resources that support the collective action.

The consideration of CCs appears to be missing in many decision-making processes concerning energy development in the Arctic. For example, during fieldwork in Northern Norway, we encountered several instances of companies attempting to develop training programmes for young people to prepare them for careers in the oil and gas industry. However, these efforts effectively limited some participants' capabilities of enjoying gainful and fulfilling employment. This was because the companies did not have a full grasp of the circumstances in which the companies and communities operate – that young people may find it difficult to find a new job when the oil and gas industry's presence in the Arctic shrinks due to the impact of energy transition policies.

Building a blueprint to develop knowledge and developing individual and collective capabilities within Arctic communities to achieve justice needs a different understanding of community (*northern society*) as *a relational experience* in which all activities are bound together with all their differences and conflicts that are mutually constitutive. This experience is also structured at different scales and dimensions, from being a member of a local community to a citizen of the Arctic and from representing a particular industry to serving on a school board. Seeing the good life as a relational experience and grasping all the different aspects, conditions, conflicts and interrelations presents a challenge for those with decision-making responsibilities. This challenge manifests itself in how oil and gas companies provide support for Arctic Indigenous communities (reindeer herders for example). These activities often range from supplying day-to-day goods and services to building schools and sports centres for their relatives living in settlements. However, there has been limited or no investment in what is more significant for the reindeer herders, like infrastructure to support traditional economic activity – for example a centre where deer meat could be delivered, deer meat processing workshops, or shops directly in the place where they live.

Capabilities and energy justice

Energy justice is a rapidly developing concept that has yet to be fully integrated with more established schools of justice. Surprisingly, the CA is not an exception, given the applied nature of both concepts. However, there have been several attempts to connect it to energy justice. For example, Benjamin Sovacool and Michael Dworkin (2014, p 437), in their attempt to conceptualize energy justice, note that energy poverty 'interferes with human beings' ability to achieve functions and capabilities'. Kirsten Jenkins et al (2016) underscore the impact of energy development on the capabilities of Indigenous communities. Benjamin R. Jones et al (2015, pp 151–60) connect their energy justice framework and the CA in a more direct fashion. They do so by building the following four foundational assumptions:

1. Every human being is entitled to the minimum of basic goods of life that is still consistent with respect for human dignity.
2. The basic goods to which every person is entitled also include the opportunity to develop the characteristically human capacities needed for a flourishing human life.
3. Energy is only an instrumental good – it is not an end in itself.
4. Energy is a material prerequisite for many of the basic goods to which people are entitled (Sovacool et al, 2014).

Based on these assumptions they arrive at the following principles, (1) prohibitive and (2) affirmative:

1. Energy systems must be designed and constructed in such a way that they do not unduly interfere with the ability of any person to acquire those basic goods to which he or she is justly entitled.
2. If any of the basic goods to which every person is justly entitled can only be secured using energy services, then in that case there is also a derivative right to the energy service.

These two principles serve as the positive and negative limits of when energy services *must be available* and how energy services *cannot be procured*. Thus, to make the CA work within this two-principle framework, one must connect the principles to the assumptions two and four. In the second assumption, Jones et al (2014) adopt Nussbaum's and Sen's rejection of the sufficiency of basic goods in defining human well-being. In the fourth assumption, they recognize the instrumentality of energy for developing human capabilities.

Bethel Tarekegne and Roman Sidortsov (2021) develop recognition, distributive, procedural and restorative guiding principles for expanding electricity access in sub-Saharan Africa by fusing the affirmative principle (AP) and prohibitive principle (PP) and the CA. However, they do not restate the principles as based on capabilities instead of basic goods. In addition, they do not elaborate on the individual and collective capabilities as part of the CA. We rectify these shortcomings below:

PP: energy systems must be designed and constructed in such a way that they do not unduly interfere with individual and/or collective capabilities.

AP: energy services must be provided if they are instrumental to securing individual and/or collective capabilities but only in a manner consistent with the prohibitive principle (through energy systems that are designed and constructed in such a way that they do not unduly interfere with individual and/or collective capabilities).

Collective capabilities and energy justice in the Arctic oil and gas development context

The energy transition and the economic, social and environmental challenges posed by the rapidly warming Arctic involve a wide range set of CCs. By energy being an instrumental good, these CCs impact other individual and collective capabilities necessary for human and societal flourishing. For example, if a community has sufficient financial resources to mitigate the loss of local oil and gas jobs, the community can protect its members, reorient itself in the changed circumstances, and build on its identity. We propose an analytical framework (Table 5.1) that categorizes these CCs as (1) *recognizing* the need of political and social will for the energy transition, (2) *enduring* the impacts of the socio-economic realignment and restructuring due to the energy transition, and (3) *transforming* the socio-technical systems to flourish during and post the energy transition. These three categories are premised on the fact that the energy transition is a complex, difficult, and controversial process that resembles more than just a switch from one technology to another (Meadowcroft, 2009). It involves a plethora of social and political mechanisms and rules working together to transform the existing energy system away from fossil fuels.

Most if not all energy transition scholars agree that the new energy system should be decarbonized and sustainable (Meadowcroft, 2009; Sovacool, 2016). As a result, the need to mitigate the impacts of climate change looms large as the dominant driver behind the energy transition.[1] Because of the pervasiveness and severity of the threat to the human

---

[1] We recognize the concept of just transition that focuses on the well-being of the workers impacted by the energy transition.

Table 5.1: Analytical framework for assessing the energy transition – the collective capabilities and energy justice perspective

CC type	CC examples in the context of PP and AP
Recognizing	*Procedural CC* PP: Financial support and procedural opportunities for public intervention in climate lawsuits AP: Holistic energy planning processes that account for the declining costs of renewable energy *Substantive CC* PP: Adequate public funding for climate science research AP: Public energy education, outreach and awareness
Enduring	*Procedural CC* PP: Public oversight overextensions and improvements of oil and gas facilities AP: Simplified procedure for permitting distributed renewable energy projects *Substantive CC* PP: Oil and gas industry-funded decommissioning funds AP: Energy-efficiency programmes for winterization of housing stock
Transforming	*Procedural CC* PP & AP: Procedural opportunities for children above a certain age to participate in constructing visions of their community's future *Substantive CC* PP: Internal combustion engine (ICE) bans AP: Oil and gas industry-funded programmes targeting decarbonization of cement and steel production

civilization, concerns like maintaining traditional ways of life and human development in disadvantaged communities (both are numerous and critical in the Arctic) are often seen through the lens of climate change impacts. The disappearing coastal plain, destroyed and abandoned villages, and the displaced people who inhabited these places for generations are examples of *why* the energy transition is necessary (Shearer, 2012). However, the energy transition can also be a contributing *cause* of disappearing traditional ways of life and lagging human development in the Arctic, for example, when a wind energy project interferes with reindeer herding or when a diesel generator remains the only viable option for powering a local school or hospital. Therefore, it is mandatory to balance the necessity of energy for developing capabilities with the negative impacts that energy has on such development. The wind farm can be sited offshore, and the diesel generator can be replaced when a hybrid (renewable energy generation coupled with batteries) is installed. This balance is the intended practical outcome of the AP and PP and the reason for which we included them in the proposed analytical framework.

As we previously noted, it is critically important for a society to have the freedom to engage in public discourse and facilitate the formation of both individual and collective capabilities. Therefore, we add further nuance to the proposed framework by distinguishing procedural and substantive CCs within each of the three categories. However, it is doubtful that such public discourse will be effective if it lacks a financial, institutional and knowledge foundation that serves collective interests and not individual ones. The efficacy of a comprehensive, inclusive and the otherwise well-designed process will be severely reduced if a collective body does not have the knowledge and organizational, institutional and financial capacities to support the discourse or to act on its outcomes.

The success of the energy transition is likely to depend not on technological breakthroughs but on how well the socio-political mechanisms through which the transition occurs can overcome the vested interests created by the fossil fuel status quo. These vested interests are stronger, often in terms of the overall strength and scale, in places where the status quo is the main, if not the only, economic activity. By being a global energy backyard for over a hundred years, the Arctic is one such place (Sidortsov, 2016). Therefore, the CCs that are at the forefront of challenging the fossil fuel status quo are those that are aimed at recognizing the need and developing the socio-political will for the energy transition.

Challenging the status quo begins with societies having public processes, mechanisms and forums for *recognizing* the need for doing so, as well as being able to participate in the mechanisms and access the forums. This involves scrutinizing the current system and its negative impacts that interfere with CCs. It also involves developing a new system that can support CCs at a given scale while avoiding interference. An example of the procedural PP CCs includes the ability to launch a legal challenge against the oil and gas industry based on its contribution to climate change. An instance for the procedural AP CC would be an inclusive and transparent energy planning process that assesses and evaluates the technological options and scenarios available as a result of the energy transition. Yet a court challenge or public intervention in energy planning is unlikely to succeed in the absence of unbiased scientific climate change knowledge and a basic understanding of energy systems among the general public. This is possible through publicly funded scientific research, education, outreach and awareness, which are the examples of PP and AP CCs that we note in Table 5.1.

Whereas recognizing the need and socio-political will to move away from fossil fuels is the initial step on the road to the energy transition, *enduring* the growing pains of the energy transition while *transforming* the socio-technical systems are concurrent and interdependent processes. Yet they are sufficiently distinct to form two separate categories with *transforming* CCs primarily responsible for moving societies from the current energy system to a new one while *enduring* CCs ensuring that the transition withstands the shocks of the socio-economic realignment and restructuring but ultimately stays the course.

The increasing demand and rising energy prices due to the post-COVID 19 economic recovery exposed the fragility of maintaining the energy transition momentum gained after the adoption of the Paris Agreement. The momentum has been further dampened by the decrease in oil and gas supply due to Russia's aggressive invasion of Ukraine, with growing calls for new oil and gas infrastructure and development (IEA, 2022). This crisis is not the last unexpected obstacle to the energy transition – with large, centralized energy assets such as oil and gas fields and pipelines coming offline and large energy producers reducing their oil and gas operations, the sudden drops in supply are unlikely to synch with the growing supply from low-carbon sources. Therefore, it is critical to have procedural CCs that can ensure that losses in the supply of oil and gas are not compensated with new projects. We use the example of public oversight over extensions and improvements of existing oil and gas facilities as a procedural PP CC, as these kinds of projects de facto constitute greenfield development and can become politically popular when oil and gas prices soar. To ensure that the loss in hydrocarbon supply, natural gas in particular, is met with the infusion of low-carbon sources, the energy transition needs simplified and streamlined permitting and siting procedures for low-impact renewable energy projects such as distributed solar and storage. We use this example for enduring procedural AP CCs. Having sufficient substantive CCs to safeguard against negative consequences created by the shrinkage in the oil and gas sector is the primary goal of PP CCs. Our example draws attention to sufficient funds to pay for the decommissioning of oil and gas facilities to avoid placing this costly burden on the public. Having sufficient substantive CCs ensures that the loss in oil and gas supply does not impact the provision of energy services. Energy-efficiency programmes can help to partially offset the need for heating and, as a result, lessen the dependence on natural gas that is often used to provide this energy service (IEA, 2022).

The energy transition will not happen overnight; it is likely to take decades, especially in places that are reliant on oil and gas for their economy and/or for the provisions of energy services. Most Arctic communities belong to at least the 'or' category and many have their economies and furnaces fuelled by oil or gas. Therefore, pathways towards transformed energy systems at community, subnational and national levels must be carefully thought through. They should include considerations of what is expected from the energy transition to develop individual and collective capabilities and what kind of energy transition-related interferences need to be avoided. Additionally, these pathways must be produced with the input from a wide range of the population, including those who might be too young to participate in the formal political process. Hence, we refer to procedural opportunities for children above a certain age to participate in constructing visions of their community's future as an example of transforming procedural PP and AP CCs. Transforming the energy system is not just a matter of influencing the demand side; because the abundant and affordable supply of oil and gas can override the constraints placed on demand, the ability to target both is likely to be more effective. For this reason, we note the ability to impose internal combustion engine bans and financial capacities to transform cement and steel production, which are among the hardest sectors to decarbonize, as examples of substantive PP and AP CCs.

Conclusion

The energy transition in the Arctic is likely to encounter major headwinds due to the region's reliance on the oil and gas industry. In many Arctic communities, oil and gas serve as the foundation of the local economy and most, if not all, Arctic communities rely on oil and/or gas for the provision of critically important energy services. To counter these headwinds, we propose an innovative approach that combines the energy justice and collective capabilities concepts. Our approach is premised on the notion that although public and private entities might both benefit from oil and gas projects short- and medium-term, their strategic, long-term interests and objectives diverge. Whereas the oil company's interest in a nation, province and/or local community is limited by the economic life of a hydrocarbon deposit, the interest of this nation, province and/or local community in the flourishing of its people must not.

We build our approach on collective capabilities, the procedural and substantive shared freedoms that can bind individual actions through interactions and connections into a collective one and provide resources that support the collective action. To contextualize our approach to the realities of the energy transition we employ two energy justice principles, affirmative and prohibitive, proposed by Benjamin Sovacool et al (2014). The outcome of the proposed approach is an analytical framework that categorizes the CCs needed for the energy transition into the following three groups: (1) recognizing the need of political and social will, (2) enduring the impacts of the socio-economic realignment and restructuring, and (3) transforming the socio-technical systems to flourish. We hope that the proposed framework will help researchers and private and public decision makers to conduct holistic analyses of societal capacities necessary for transitioning to a sustainable and just energy future.

Study questions

1. Why do the strategic, long-term interests of the oil and gas industry and those of national, subnational and local societies diverge?
2. What are collective capabilities and how are they different from individual capabilities?
3. What is the main difference between the PP and AP?
4. Please provide an example of a procedural or substantive collective capability under the proposed analytical framework and explain to which energy justice principle it belongs.

Acknowledgements

This chapter has received funding from the European Union's Horizon 2020 research and innovation programme under grant agreement No 869327.

References

Dewey, J. (1929) *The Later Works of John Dewey, Volume 4, 1925–1953: The Quest for Certainty*, edited by J. Boydston, Carbondale: Southern Illinois University Press.

Evans, P. (2002) 'Collective capabilities, culture, and Amartya Sen's *Development as Freedom*', *Studies in Comparative International Development*, 37(2): 54–60. https://doi.org/10.1007/BF02686261.

Fjaertoft, D., and L.P. Lunden (2015) 'Russian petroleum tax policy – continuous maneuvering in rocky waters', *Energy Policy*, 87: 553–61. https://doi.org/10.1016/j.enpol.2015.09.042

IEA (2021a) *Key World Energy Statistics 2021 – Drivers of CO2 emissions*, [online], Available from: https://www.iea.org/reports/greenhouse-gas-emissions-from-energy-overview/drivers-of-co2-emissions [Accessed 20 July 2022].

IEA (2021b) *Key World Energy Statistics 2021 – World total energy supply by source*, [online], Available from: https://www.iea.org/reports/key-world-energy-statistics-2021/supply [Accessed 20 July 2022].

IEA (2022) *A 10-Point Plan to Reduce the European Union's Reliance on Russian Natural Gas – Analysis*, [online], Available from: https://www.iea.org/reports/a-10-point-plan-to-reduce-the-european-unions-reliance-on-russian-natural-gas [Accessed 20 July 2022].

Jenkins, K., D. McCauley, R. Heffron, H. Stephan, and R. Rehner (2016) 'Energy justice: a conceptual review', *Energy Research & Social Science*, 11: 174–82. https://doi.org/10.1016/j.erss.2015.10.004.

Jones, B.R., B.K. Sovacool, R.V. Sidortsov, and the Center for Environmental Philosophy, the University of North Texas (2015) 'Making the ethical and philosophical case for "energy justice"', *Environmental Ethics*, 37(2): 145–68. https://doi.org/10.5840/enviroethics201537215.

Massingham, P. (2019) 'An Aristotelian interpretation of practical wisdom: the case of retirees', *Palgrave Communications*, 5(1): 123. https://doi.org/10.1057/s41599-019-0331-9.

Meadowcroft, J. (2009) 'What about the politics? Sustainable development, transition management, and long term energy transitions', *Policy Sciences*, 42(4): 323–40. https://doi.org/10.1007/s11077-009-9097-z.

Nussbaum, M. (2002) 'Capabilities and social justice', *International Studies Review*, 4(2): 123–35.

Nussbaum, M.C. (1995) 'Aristotle on human nature and the foundations of ethics', in J.E.J. Altham and R. Harrison (eds) *World, Mind, and Ethics*, Cambridge: Cambridge University Press, pp 86–131. https://doi.org/10.1017/CBO9780511621086.007

Robeyns, I. (2005) 'Selecting capabilities for quality of life measurement', *Social Indicators Research*, 74(1): 191–215.

Sen, A. (1985) *Commodities and Capabilities*. Amsterdam: North-Holland.

Sen, A. (1992) *Inequality Reexamined*, Harvard, MA: Harvard University Press.

Sen, A. (1999) *Development as Freedom*, Oxford: Oxford University Press.

Shearer, C. (2012) 'The political ecology of climate adaptation assistance: Alaska Natives, displacement, and relocation', *Journal of Political Ecology*, 19(1). https://doi.org/10.2458/v19i1.21725.

Sidortsov, R. (2012) 'Measuring our investment in the carbon status quo: case study of the new oil development in the Russian Arctic', *Vermont Journal of Environmental Law*, 13, [online], Available from: https://papers.ssrn.com/sol3/papers.cfm?abstract_id=2494433 [Accessed 20 July 2022].

Sidortsov, R. (2016) 'A perfect moment during imperfect times: Arctic energy research in a low-carbon era', *Energy Research & Social Science*, 16: 1–7. https://doi.org/10.1016/j.erss.2016.03.023.

Sidortsov, R., and E. Gavrilina (2018) 'When foundation matters: overcoming legal and regulatory barriers to oil and gas well decommissioning in Russia', *The Journal of World Energy Law & Business*, 11(3): 209–19. https://doi.org/10.1093/jwelb/jwy012.

Sovacool, B.K. (2016) 'How long will it take? Conceptualizing the temporal dynamics of energy transitions', *Energy Research & Social Science*, 13: 202–15. https://doi.org/10.1016/j.erss.2015.12.020

Sovacool, B.K., and M.H. Dworkin. (2014) *Global Energy Justice: Problems, Principles, and Practices*, Cambridge: Cambridge University Press.

Sovacool, B.K., R.V. Sidortsov, and B.R. Jones (2014) *Energy Security, Equality and Justice*, Abingdon: Routledge.

Tarekegne, B., and R. Sidortsov (2021) 'Evaluating sub-Saharan Africa's electrification progress: guiding principles for pro-poor strategies', *Energy Research & Social Science*, 75: 102045. https://doi.org/10.1016/j.erss.2021.102045.

Vizhina, I.A., A.A. Kin, and V.N. Kharitonova (2013) 'Problems of public-private partnership in implementation of strategic projects of the North', *Regional Research of Russia*, 3(1): 103–12. https://doi.org/10.1134/S2079970513010139.

Wood-Donnelly, C., and M.P. Bartels (2022) 'Science diplomacy in the Arctic: contributions of the USGS to policy discourse and impact on governance', *Polar Record*, 58. https://doi.org/10.1017/S0032247422000134.

6

Mainstreaming Environmental Justice? Right to the Landscape in Northern Sweden

Tom Mels

Introduction

Not everyone suffers equally from environmental degradation, nor does everyone equally enjoy environmental benefits. These inequalities clamour for attention from justice activists and scholars. What are the origins of these patterns of distribution? Why do they persist? The search for answers yields further discoveries about inequalities in the social fabric of society. It makes all the difference in the world if you belong to either the privileged or marginalized communities of justice. Exclusion from participation in social life and decision making constrains some people's control over the environment, while endowing others with benefits. Such questioning may reveal that disproportionate exposure to environmental risks emanates from structural misrecognition of the status and rights of marginalized communities. Research within the field of environmental justice seeks to analyse such environmental inequalities and the claims to justice they engender.

Nowadays, attention to environmental justice is no longer limited to activism and academic research. The importance of considering justice is increasingly recognized in mainstream environmental politics. But exactly what does this adoption entail? Juxtaposing justice conflicts over mining, energy, forestry and nature conservation landscapes in the far north of Sweden, this chapter focuses on a region increasingly appropriated by international media as 'a green jobs Klondike' (*The Guardian*, 2021). More accurately understood, current resource mobilization brings about a version of environmental justice that, in strategic ways, abstracts from the ongoing production of a distinctively capitalist biophysical and ideological landscape. This abstraction from the landscape as a peopled polity and place reproduces rather than resolves structural injustice, including imperilling Indigenous livelihoods. In contrast to these mainstreaming tendencies, the chapter envisions environmental justice as a critical normative engagement with multiple contested boundaries of nature's commodification in the course of capitalist development.

Justice enthusiasm

Over the course of a few decades, there has been a steadily growing academic interest in understanding how social justice and environmental issues intertwine (Coolsaet, 2020). Nowadays, justice appears virtually everywhere in environmental policy making and planning discourse too, encompassing the pedestrian level of cities and countryside as much as the lofty realm of global conventions. Whenever there is an environmental problem calling for action, it is almost automatically accompanied by the language of justice: climate change comes with climate justice; biodiversity loss must include species justice; sustainable agriculture cannot be conceived without agricultural justice; organic food provision ushers in food justice; decarbonized energy production associates with energy justice; urban greening asks for considerations of environmental gentrification; transport system development leads to discussions on transport justice.

Behind this emergent profusion of justice foci lies a question-begging array of positions. In policy making and planning, justice is in many respects adapted to largely mainstream forms of sustainable development. On a planetary scale, the Global Goals (United Nations, 2015) embrace a host of *distributive* justice principles, including benefit sharing and access to economic resources and public spaces, several pleas to solve *recognition* conflicts, such as those related to gender inequality, and calls to *procedural* justice ensuring 'responsive, inclusive, participatory and representative decision-making at all levels' (Target 16.7). On a continental scale, the European Commission's implementation strategy of such global goals, the European Green Deal (2019) speaks of a 'mainstreaming of sustainability in all EU policies', staging a vaguely defined 'just transition', heavily resting on investment plans and other, standard economic manoeuvres and measurements (European Commission, 2019, pp 15–16). Accompanying this pleonastic logic of the mainstreaming of sustainability in the EU thus comes a mainstreaming of justice too that strategically fails to challenge the many underlying causes of oppression and ecological ruin in capitalist society. Adding more calls to justice into the standard sustainability mix of providing economic, ecological and equity benefits thus hardly seems to instil much political panic these days. It may even be seen as an innocuous add-on to the often equally bland language of sustainability.

Contrary to such a reading, it can be argued that the newfangled appeals to justice should be celebrated, simply because erstwhile marginalized notions of justice are gradually becoming the centre of attention. For whatever referential vagueness and practical futility, the recent turn to justice provides a language that, in a further process of substantiation, may oblige mainstream enthusiasts to consider socio-economic maldistribution, make procedural demands to political representation, or challenge cultural misrecognition, and even pay heed to lengthy histories of social struggle and inequality. Over the past decades, a rather extensive academic corpus suggests as much, including scholarly work on *just sustainabilities*. Empirical evidence shows it is possible to alleviate what Julian Agyeman (2005, p 44) calls the 'equity deficit' of mainstream sustainable development, pervading 'most "green" and "environmental" sustainability theory, rhetoric, and practice' (Agyeman, 2013, p 4). However, as Agyeman readily acknowledges, 'there are strong forces ranged against such change: wealthy elites, corporate interests, and governments playing "race to the bottom" to attract inward investment and maintain reckless economic growth' (Agyeman, 2013, p 164). The extensive turn to a greening of policies and practices motivated by a plethora of just sustainability transitions needs accordingly to address the underlying societal dynamics that drive that race to the bottom. In tracing that race, 'all roads lead to one idea – namely, capitalism' (Fraser, 2021, p 96).

My tentative hypothesis is that the current justice enthusiasm also entails a development towards what could be described as a *mainstreaming* of environmental justice. The diagnosis is that in the movement from civic, on the ground activism and the overt indignation of marginalized communities to the (avowedly) green fervours of corporations, bureaucracies and policy makers, something has been lost. To explain this loss, the proliferation of greening and justice needs to be scrutinized within the broader 'institutionalized social order' (Fraser, 2014a) that is capitalist society. The tendency to mainstreaming entails that claims to environmental justice are circumscribed and narrowed down to ambitions and practices adapted to, and hence fundamentally unable to challenge, that order. I will illustrate this by briefly exploring northern Swedish landscapes in crisis. After all, capitalism is not just an institutionalized social order; it is also a non-accidental spatial order dependent on the landscape that makes that institutionalized order possible.

Landscapes under pressure

The mainstreaming of justice needs to be grounded in the material landscape where that institutionalized social order takes shape. In northern Sweden, nature as the raw material for the timber industry and a source for energy provision, alongside mining for ores and metals, provides an important context. These illustrate empirically central features of capitalist society and not just because they are evidently vital to commodity production for profit.

Taking a cue from Nancy Fraser's work, they reveal, first, how commodity production is something that relies on a supporting surrounding. Put more theoretically, the economic 'foreground' of commodity production depends on non-economic enabling conditions such as nature (minerals, forests, energy resources), and public power (planning and legislation concerning mining, forest management, energy transition). They also depend on the sphere of social reproduction unfolding in the everyday social lives and experiences of communities and workers. Second, they show that in 'a system devoted to the limitless expansion and private appropriation of surplus value' these non-economic enabling conditions tend to be increasingly commodified (Fraser, 2018, p 5). Third, this process of commodification is highly uneven. Fraser claims it entails the forced expropriation and exploitation of capacities and resources, the political imposition of uneven status hierarchies, and social domination. Fourth, because of this unevenness, processes of commodification do not go unchallenged. They cause a host of 'boundary struggles' in which opposing claims to justice come forth. The landscape of northern Sweden will show what this all means in practical terms.

Mining as a moral duty

Mineral resources occupy a special nook in Swedish sustainability discourse and policy making (Tarras-Wahlberg and Southalan, 2021). Recently, the Swedish Minister for Business, Industry and Innovation, Karl-Petter Thorwaldsson, claimed it is 'almost a moral duty to open new mines' in the country to secure a strategic supply of (critical) raw materials (*Sveriges Natur*, 2022). This moral call fits well into the notoriously neoliberal Swedish Mineral Law (SFS, 1991) and the concomitant rhetoric of the Swedish Mineral Strategy (Näringsdepartementet, 2013).

Public efforts to facilitate mining concessions, prepare necessary physical infrastructure and offer ready access to the archives of the Swedish Geological Survey all invite capital to roam the country for mineral resources. They are also attuned to political discourse on expanding the exploitation of mineral resources within the boundaries of the European Union (European Commission, 2008).

Confirming Fraser's point about the capitalist economy's reliance on public power, these also bring about justice issues. On the one hand, EU political discourse reveals an awareness of social and environmental challenges such as the 'risks of human rights infringements … or environmental destruction' (EESC, 2021, 5.9) and 'the lack of public acceptance for mining in Europe' (European Commission, 2020, p 14), yet in the hierarchical order of other potential moral responsibilities, the duty to mining relies on sharp geographical confines: 'mining in Europe is operating at the highest environmental and social standards compared to non-EU countries' (EESC, 2021, 5.14). Outside Europe, mining is often plagued by 'social exploitation and environmental pollution with usually only a few profiteers' (EESC, 2021, 5.15). This, in addition to the various green growth policies, motivates proposals about the development of 'a streamlined authorization process for mining activities' in the EU (EESC, 2021, 1.5). The European Commission's policy considerations in Critical Raw Materials Resilience (2020) also resulted in the instant launching of the industry-driven European Raw Materials Alliance (ERMA), with its outright aim of promoting public acceptance of the role of these materials in the green transition.

The notion of promoting public acceptance arguably underestimates the contradictions facing green transition policies, boundary struggles over mining in the landscape, and the tenacity of appurtenant conflicting claims to justice. As the largest producer of iron ore in the EU, and a leading exporter of copper, zinc lead, gold and silver, Sweden has substantial economic interests in developing its mineral deposits. The northern counties are particularly rich in minerals and host plenty of mining operations. From an environmental justice point of view, knowing that green growth spells exploitation, these also remain highly controversial.

In recent years, this has been very clear from the lengthy conflict over the unexplored Gállok/Kallak iron ore deposit in Jokkmokk municipality, beginning with an application for an exploitation concession permit by Jokkmokk Iron Mines AB (JIMAB, subsidiary of the UK based Beowulf Mining PLC) in 2013. Deeply troubled by ongoing deliberations, Håkan Jonsson, Chairman of the board of the Sámi Parliament, criticized the Swedish government's continued prioritization of mining interests over those of reindeer herding and Sámi culture. This practice countered a number of international guiding principles including the Universal Declaration of Human Rights, the UN Declaration on the Rights of Indigenous Peoples and the Convention on Biological Diversity. Following widespread protests, the County of Norrbotten refused to issue mining permits in 2014. However, this decision was overruled by the Mining Inspectorate of Sweden, which transferred decision making on the case to the Swedish Government (Länsstyrelsen Norrbotten, 2017). By March 2022, the firm received an exploitation concession, allowing for further environmental and economic enquiries in the mining landscape.

Critics of mining in the area (located within the Laponia World Heritage Site) point out that green transition arguments are being mobilized for the promotion of fossil-free ironworks and access to critical minerals, while underestimating environmental risks, including drinking water contamination and Indigenous rights. Longstanding controversies such as these do not just bear witness to a continued threat to local communities' right to the landscape, but, as Jonsson notes, to the fact that 'green arguments are driven forward for further exploitation of the Sámi heartland' (*Aktuell Hållbarhet*, 2021).

Another recent case concerns exploitation concession applications by Nickel Mountain AB for Rönnbäck/Rönnbäcken in Storuman municipality, which would be incompatible with the traditional pasture rotation for reindeer. By the end of 2020, the UN Committee on the Elimination of Racial Discrimination (CERD, 2020) concluded that the plans for mining were discriminatory by virtue of procedural justice deficiencies. While the CERD acknowledged the role of mining as a legitimate public interest, it criticized the absence of dialogue and consultation in a situation where the local Sámi community – not all of whom are reindeer herders – was under severe psychological pressure over threats to their livelihood. As a result, combined with profitability issues, the project was eventually shelved.

It's electrifying

As the green transition discourse suggests, developments in mining are immediately implicated in sustainable energy transition. With the emergence of a Nordic 'battery belt', with factories in Skellefteå (Sweden), Mo I Rana (Norway) and Vasa (Finland), it is unlikely that arguments about moral duties and securing supplies, combined with the sway of promoting public acceptance of mining, will cease anytime soon. Given such developments, reinvigorated attention to the nickel, cobalt and magnetite supplies at Rönnbäck, and minerals in other places, is to be expected.

While the energy-intensive metals and mining sector in Sweden traditionally relied almost exclusively on fossil fuels, the authorities have now settled for a national electrification strategy (Regeringskansliet, 2022). In 2016, the Swedish Parliament adopted a long-term energy policy aiming for a fast expansion of renewable electricity generation. With hydropower already developed on an industrial scale in earlier rounds of energy planning in the north, the current priority is a considerable expansion of wind power. Once fully developed, Svevind's Markbygden 1101 west of Piteå will become Europe's largest on-land wind farm, expected to generate around 8 to 12 TWh per year – a substantial part of Sweden's planned wind power (Energimyndigheten, 2021).

While Markbygden has created only limited controversy, this is not the case with furniture giant Ikea's wind power development in the mountains of Glötesvålen, Härjedalen. The project has attracted ample scholarly attention and featured in the well-known Swedish Radio series *Konflikt* (SR, 2021; Skarin and Alam, 2021). After a lengthy struggle with private landowners over Sámi rights to herding in the 1990s, a curiously biased environmental impact assessment supported commercial wind power development in the region.

Like the hydropower projects of the twentieth century (for example, their impact on hydrology and landscape ecology in places like Rönnbäck), wind turbines stir controversy. Legal scholars have scrutinized inbuilt procedural justice issues of standard planning practices, where the state relinquishes consultation responsibilities to project developers. These practices demonstrate the state's failure to recognize the status of the Sámi reindeer herders as Indigenous people, including their special rights to land and resources (Allard, 2018; Cambou, 2020). They inspire one-way information exercises imbued with a belief in the viability of win-win solutions and the harmonious co-existence of reindeer herding and energy industries in the landscape. They also sustain a narrow *stakeholder* perspective, where reindeer herders are considered with regards to their entrepreneurial interests, and hence made fully comparable to wind power companies or other businesses, rather than special rights-holders concerned about their livelihoods (Larsen and Raitio, 2019, p 15; Bjärstig et al, 2020, p 13). Essentially, legal and policy arrangements structurally codify and enforce these matters in terms of a commodity logic that to an extent succeeds in escaping scrutiny in a wider, politicized justice perspective. What the resulting boundary struggles show, however, is that these livelihoods and rights embody social practices and values that clearly surpass commodity logic (see, for instance, Fraser, 2014a, pp 66, 69).

These examples of mining as a 'moral duty' and the electrification strategy show, in the first place, that any 'romantic view' that construes nature (mining resources, material landscape) and polity as intrinsically separate from capitalism simply is misguided (Fraser, 2014a, pp 69–70). Yet Fraser claims this separation is exactly what capitalist society's 'normative topography' tends to institutionalize (Fraser, 2014a, p 67). In the mining and energy field, the discussion seems to be about solving environmental crises by providing resources for sustainable transitions. It installs 'a natural realm, conceived as offering a free, unproduced supply of "raw material" that is available for appropriation' (Fraser, 2014a, p 63). Questioning the inner workings and commodity logic of capitalist society is not part of the equation. How commodification colonizes virtually all aspects of social life, sustaining exploitative processes and social dominance, hence is obfuscated (Fraser, 2018, p 3). Under these circumstances, it is hard to mobilize justice as a critical concept to grasp the broader institutionalized order that is capitalist society.

In the second place, mining and energy show that with every new boundary struggle, capitalist society breeds additional fronts in which it can become vulnerable. Capitalist environmental practice will thereby also continue to threaten the social and ecological conditions that render accumulation at all possible, inevitably giving rise to new forms of crisis (Fraser, 2014a, p 63).

Right to the forest

In mining and energy landscapes, at the boundaries of commodification, multiple crises occur, accompanied by numerous forms of social resistance. This crisis tendency may call into existence new social movements questioning capitalist society's normative topography (Fraser, 2014a, p 69) and capitalism's subsumption of nature (see, for instance, Holifield et al, 2018). Regardless of their content, the frontiers of commodification will thus inevitably engender different claims to justice by different communities. A case in point is the so-called Forest Revolt (*Skogsupproret*), initiated by Sámi activists and environmentalists. Militating against 'Sweden's colonial forest destruction', the protesters made calls to 'decolonize Sápmi' (that is, the space traditionally inhabited by the Sámi people) and to 'democratize the forest' (Skogsupproret, 2022).

First, the identification of logging with a *colonial* present emanates from continued claims to injustice, in particular regarding Sámi rights. Granted, over the past decades, a number of significant legal and political changes have been made: Swedification policies were aborted in the late 1970s; the Swedish state constitutionally recognized the Sámi as an Indigenous people in 2010; and in 2021 a truth commission was set up to examine historical injustices levelled against the Sámi (Regeringen, 2021). The Swedish Supreme Court's recent (2020) recognition of exclusive hunting and fishing rights of the Sámi reindeer herders of Girjas has also been identified as 'an important victory' for the community by asserting its customary rights to land and natural resources. However, attention to the historical utilization of the landscape also roused 'increased racism and conflict between groups of Sami as well as between Sami and Swedish locals' (Allard and Brännström, 2021, pp 56, 57; see, for instance, Dahre, 2004). Moreover, it remains to be seen if these legal developments – the more sustained consideration of the ancient Swedish property law concept of immemorial prescription (*urminnes hävd*) and usufructuary rights to herd reindeer on public and private land – will prove to be of decisive aid in solving conflicts and remaining disagreements over the right to the landscape in Sápmi.

Second, the activists' demand to *democratize* the forest arguably emanated from the protracted historical experience of expropriation (central to Indigenous rights struggles), and it ties in with the increasingly vibrant rights agenda in international forestry. The latter encompasses a diverse repertoire of rights claims for the redistribution of benefits, the recognition of forest people's identities, and active promotion of participatory justice (see, for instance, Sikor and Stahl, 2012). Deficiencies in procedural rights, such as opportunities to participate in environmental decision making regarding the forest, were a key issue behind the blockade against state-owned Sveaskog, in Jouksuvaara, Pajala municipality in May 2021. In a letter to the editors of the *Aftonbladet* newspaper, 29 Sámi communities summarized widespread grievances about the consequences of the ongoing privatization of state-owned forests (a programme initiated in 2002 by the then conservative government to support private forestry and regional development, in particular in the far north; see Sveriges Riksdag Näringsutskottet, 2020):

> According to the State's owner policy, state companies should set a good example. But instead of being a paragon for the industry, Sveaskog seems to use the land sales program from 2002 as a pretext to systematically sell those very lands on which the Sámi communities year after year prevented logging precisely because they are vital to reindeer husbandry. The private landowner who acquires the property will usually be exempted from consultation obligations and, because supervision remains inadequate, will in principle be free to harvest the forest. After the felling, or as part of the business agreement, Sveaskog can buy back the timber from the private forest owner. By the time the Sámi community becomes aware of the deal, the land is usually already bare-cut and lost. The forest companies are destroying our reindeer pastures. (*Aftonbladet*, 2020)

This is certainly not a novel phenomenon and appears to reflect the legal prioritization of timber production over reindeer herding (Brännström, 2017). Over the past few decades, private landowners and the forest industry continue to assail Sámi use rights, based on controversial claims regarding property values and biodiversity:

> The right of use is now being challenged in many places. Private landowners are literally conducting a legal hunt against the Sami today to inhibit reindeer herding. These legal cases are not just about financial compensation for damage to the forest. More fundamentally, they are about the existence of the Sami. (Dahre, 2004, p 454)

Such regional skirmishes provoke international indignation over Indigenous rights and environmental conduct too. Relying on existing legal provisions, Sweden has resisted ratification of the International Labour Organization's 1989 Indigenous and Tribal Peoples Convention (ILO Convention No. 169) with its demands on the procedural right to consultation and recognition of land rights (the latter included Sámi reindeer husbandry and small game hunting and fishing).

This reverberates the justice claims surrounding mining and energy. Public investigation has made abundantly clear what is at stake: a fear of jeopardizing state and capitalist control over vast tracts of a resource-rich landscape (SOU, 1999, p 25). In the meantime, continued international discontent expressed by the UN, the Council of Europe and human rights bodies suggests that Swedish legislation does not meet the internationally accepted standards regarding the protection of Indigenous rights. The persistent appeal is for Sweden 'to properly demarcate traditional Sámi land areas and adopt legislation that recognizes and protects Sámi land and resource rights, as well as secures legal aid to allow Sámi to assert their rights before the courts' (Allard and Brännström, 2021, p 60). In the meantime, too, Sveaskog – the corporatized scion of what once was state forestry and currently the largest forest owner in the country – was awarded the Swedish 'greenwash prize' 2020 by Friends of the Earth for what was perceived as its strategic hoodwinking.

Production–reproduction

All over the board, conflicts over forests, minerals and energy production, and small game hunting and fishing rights, have multiplied, and they illustrate the actuality of Fraser's notion of ongoing boundary work in capitalist development. These examples of using northern nature as a source of input to accumulative society – resulting in clearcuts, noise and air pollution, land degradation, tailings and waste dams – are familiar enough.

Although Fraser highlights important contradictions involved in treating nature as 'raw material' and 'sink' (Fraser, 2014a, p 63), she also insists that this strand of crisis needs to be connected to a larger social totality (Fraser, 2014b, p 549). The circumstances sketched above reveal how nature emerges in the sphere of reproduction, or, more accurately, in the production/reproduction nexus as identified by feminist geographers as *life's work* (see, for instance, Mitchell et al, 2012). In Sámi lived realities and the material social practice of reindeer herding, labour and capital intimately and reciprocally relate to the environmental conditions of production. In other words, Sámi reproduction of the means of production inevitably encompasses the *environmental* conditions of production. The reconfigured relations with the primary sector of forestry, mining and energy noted earlier, precisely *because* they adversely affect these environmental conditions, therefore also entail a mounting crisis in social reproduction. Such an existential diagnosis sits well with environmental justice scholarship.

In addition to this, both capitalist capture of natural resources and traditional Sámi use value are increasingly confronted by (global) interests of using the landscape as a space for commercial tourism development and recreation. This is perhaps most obviously the case in the practice of nature conservation.

As a complement to accounts of mobilizing nature as raw material, there is the equally exclusionary practice of capturing the north as non-human wilderness and terrain of social reproduction for tourists: fabricating rural remoteness as a space for replenishment of urban labour power. Codifying on a deeply racialized status hierarchy, the Sámi appeared in earlier policies and representations as either a pre-modern voiceless part of nature (hence excluded from participatory justice), or a modern threat to the wilderness experience. Current tendencies toward the commodification of nature conservation in Sweden reveal a changing frontier (Mels, 2020). Regional development driven by a global experience economy once again makes the landscape subject to confiscation on the part of capital. 'Who has the right to the mountains?' is indeed a seriously complex question in this context, explored by Swedish television in a series highlighting 'the struggle over the mountains', where increased tourist presence interferes with reindeer herding (SVT, 2021).

Conclusion: Mainstreaming justice

The cases sketched in this chapter confirm a well-known theoretical observation insisting on attention to the geographical production of inequalities as a core focus for justice scholarship. Distribution, participation and recognition are not just philosophical concepts. They combine practically and confront us in the landscape as a peopled polity and place. However, current use of the language of justice in policy making and planning seems to make little practical difference to these environments. To rephrase: What is it about capitalist society that engenders the mainstreaming of environmental justice?

Landscapes under pressure help answer that question because they show how environmental justice claims are framed to fit the fabric of capitalist society. Mainstreaming tendencies seem to emerge from an ideologically driven reluctance – tied to political and economic interests of ongoing exploitation – to get in the way of ongoing commodification (Fraser, 2021, p 100). They serve an indispensable political function for accommodating rather than structurally questioning capitalism's normative topography (Fraser, 2015). They thereby support 'the sustainability of capitalism', rather than 'the sustainability of society and nature' (Fraser, 2014b, p 549). Against such mainstreaming stands critical attention to justice issues concerning resource redistribution, cultural recognition and participation in policy development and planning (see, for instance, Fraser, 2008). All of these haunt the landscapes of forestry, mining and energy.

The boundaries of commodification cover critical historical and geographical terrain of social struggle in northern Sweden. If it is true that the 'heartland of exploitation' was to be found in the urban core, then the northern periphery constituted in that sense the 'iconic site of expropriation' (Fraser, 2018, p 7), although the two are nowadays internally articulated.

Grasping this terrain is an antidote against wistful idealism. For all of its metaphorical intention, *The Guardian's* characterization of the far north of Sweden as a new green Klondike is historically misleading because, as journalist Po Tidholm (2021) has noted, it may reproduce the unfortunate stereotype of a region 'which after the latest boom has returned to some kind of wilderness that should be exploited, built and inhabited – again'. Not unlike earlier rounds of exploitation, the current stampede is motivated by 'a higher, national, purpose: the transition to a green society'. This 'recurring image of Norrland as a virgin source of raw materials' matters precisely because of earlier historical experiences of 'an infrastructure built primarily to transport raw materials, not to connect people and communities. And these communities have not been given proper tools for the future'. Rather than dreaming up spatial and historical similes, then, it is of immediate importance to ask questions like 'For whose sake is Norrland being exploited this time?' (Tidholm, 2021).

With threats to land rights from developments dictated by the market, it is indeed not altogether surprising to find even environmentally minded critics of the current Green Deal thinking about, and talking of, green colonialism. Are they facing a substantial change in dealing with environment and justice, the mere rhetorical adaptation of run-of-the-mill sustainable development or, indeed, old-fashioned greenwashing? Do the examples sketched in this chapter reveal a concern with the sustainability of capitalism or the sustainability of society and nature?

Against variants of philosophical idealism stand the tough realities of material landscapes. Forests, mines, energy and conservation raise increasingly unsettling concerns about groups being marginalized in social life and consigned to political subjectification. They reveal the occurrence of *multiple* struggles over raw materials and the cumulative effects in various parts of the landscape, often by the same people over the course of history (Naturvårdsverket, 2020). They produce landscapes of intensified boundary work where the exploitation–expropriation nexus has all but sustained environmental injustice.

Study questions

1. What are the relationships between natural resource use and justice in the cases described?
2. Can you explain what boundary struggles are all about and why they may help if you want to grasp capitalist society?
3. What do you see as viable solutions to the environmental justice issues identified here?

Acknowledgements

This work was supported by Formas grant 2020-00036.

References

Aftonbladet (2020) *Skogsbolagen skövlar våra renbetesmarker*, Aftonbladet, 26 November.

Agyeman, J. (2005) *Sustainable Communities and the Challenge of Environmental Justice*, New York: New York University Press.

Agyeman, J. (2013) *Introducing Just Sustainabilities: Policy, Planning and Practice*, London: Zed Books.

Aktuell Hållbarhet (2021) *Sametinget: 'Gröna' argument används för att exploatera Sápmi*, 22 December.

Allard, C. (2018) 'The rationale for the duty to consult indigenous peoples', *Arctic Review on Law and Politics*, 9: 25–43.

Allard, C., and M. Brännström (2021) '*Girjas reindeer herding community v. Sweden: analysing the merits of the Girjas case*', *Arctic Review on Law and Politics*, 11: 56–79.

Bjärstig, T., V. Nygaard, J.Å. Riseth, and C. Sandström (2020) 'The institutionalisation of Sami interest in municipal comprehensive planning', *International Indigenous Policy Journal*, 11(2): 1–24.

Brännström, M. (2017) 'Skogsbruk och renskötsel på samma mark: en rättsvetenskaplig studie av äganderätten och renskötselrätten', Dissertation, Umeå: Umeå University.

Cambou, D. (2020) 'Uncovering injustices in the Green Transition', *Arctic Review on Law and Politics*, 11: 310–33.

CERD (2020) United Nations, International Convention of the Elimination of All Forms of Racial Discrimination CERD/C/102/D/54/2013, 26 November.

Coolsaet, B. (2020) *Environmental Justice: Key Issues*, London: Routledge.

Dahre, U.J. (2004) 'Rädda en varg, skjut en same: den historielösa jämlikheten och föraktet för samer i Sverige', *Nordic Journal of Human Rights*, 22(4): 448–61.

EESC (2021) Opinion of the European Economic and Social Committee on 'Communication from the Commission to the European Parliament, the Council, the European Economic and Social Committee and the Committee of the Regions – Critical Raw Materials Resilience: Charting a Path towards greater Security and Sustainability' (COM (2020) 474 final) OJ C220/118.

Energimyndigheten (2021) *Nationell strategi för en hållbar vindkraftsutbyggnad*, ER 2021:2.

European Commission (2008) The Raw Materials Initiative: Meeting our Critical Needs for Growth and Jobs in Europe (Communication) COM (699) final.

European Commission (2019) *The European Green Deal* (Communication) COM (640) final.

European Commission (2020) *Critical Raw Materials Resilience: Charting a Path towards greater Security and Sustainability* (Communication) COM (474) final.

Fraser, N. (2008) *Scales of Justice,* New York: Columbia University Press.

Fraser, N. (2014a) 'Behind Marx's hidden abode', *New Left Review*, 86: 55–72.

Fraser, N. (2014b) 'Can society be commodities all the way down?', *Economy and Society*, 43(4): 541–58.

Fraser, N. (2015) 'Legitimation crisis?', *Critical Historical Studies*, 2(2): 157–89.

Fraser, N. (2018) 'Roepke lecture in economic geography – from exploitation to expropriation', *Economic Geography*, 94(1): 1–17.

Fraser, N. (2021) 'Climates of capital', *New Left Review*, 127: 94–127.

The Guardian (2021) 'Wanted: 100,000 pioneers for a green jobs Klondike in the Arctic', *The Guardian*, 21 November.

Holifield, R., J. Chakraborty, and G. Walker (2018) *The Routledge Handbook of Environmental Justice*, London: Routledge.

Länsstyrelsen Norrbotten (2017) 'Yttrande över Regeringskansliets begäran om komplettering gällande ansökan om bearbetningskoncession för Kallak K nr. 1 i Jokkmokks kommun', Diarienummer 543–14195–2017.

Larsen, R.K., and K. Raitio (2019) 'Implementing the state duty to consult in land and resource decisions', *Arctic Review on Law and Politics*, 10: 4–23.

Mels, T. (2020) 'The deep historical geography of environmental justice', *Annales de Géographie*, 736(3): 31–54.

Mitchell, K., C. Katz, and S.A. Marston (2012) *Life's Work: Geographies of Social Reproduction*, Malden, MA: Wiley-Blackwell.

Näringsdepartementet (2013) *Sveriges mineralstrategi. För ett hållbart nyttjande av Sveriges mineraltillgångar som skapar tillväxt i hela landet* (N2013.02), Stockholm: Regeringskansliet.

Naturvårdsverket (2020) *Omtvistade landskap: Navigering mellan konkurrerande markanvändning och kumulativa effekter,* Rapport 6908, Stockholm: Naturvårdsverket.

Regeringen (2021) *Kommittédirektiv: Kartläggning och granskning av den politik som förts gentemot samerna och dess konsekvenser för det samiska folket.* Dir.2021:103.

Regeringskansliet (2022) *Nationell strategi för elektrifiering*. Stockholm: Regeringskansliet.

SFS (Swedish Code of Statutes) (1991) *Minerallag* (1991: 45), Stockholm: Näringsdepartementet.

Sikor, T., and J. Stahl (2012) *Forests and People*, London and New York: Taylor and Francis.

Skarin, A., and M. Alam (2017) 'Reindeer habitat use in relation to two small wind farms, during preconstruction, construction, and operation', *Ecology and Evolution*, 7(11): 3870–82.

Skogsupproret (2022) [online], Available from: https://www.facebook.com/skogsupproret/ [Accessed 1 January 2022].

SOU 1999:25 (1999) Utredningen om ILO:s konvention nr 169. *Samerna – Ett Ursprungsfolk i Sverige: Frågan Om Sveriges Anslutning Till ILO:S Konvention Nr 169: Betänkande*, Stockholm.

SR (2021) Vindparkskonflikten i Härjedalen. sverigesradio.se/avsnitt/repris-vindparkskonflikten- i-harjedalen-ikea-och-renskotarna, 8 August [Accessed 7 January 2022].

Sverige Riksdagen Näringsutskottet (2020) *Sveaskogs samhällsuppdrag om markförsäljning: en uppföljning*, Rapport från riksdagen 2020/21:RFR3.

Sveriges Natur (2022) 'Näringsministern ser stort behov av fler gruvor i Sverige', Sverigesnatur.org/intervju, January 21 [Accessed 5 February 2022].

SVT (2021) 'Vem har rätt till fjällen', 7 April, [online], Available from: www.svt.se/special/vem-har-ratt-till-fjallen/ [Accessed 14 February 2022].

Tarras-Wahlberg, H., and J. Southalan (2021) 'Mining and indigenous rights in Sweden', *Mineral Economics*, 8. https://doi-org.ezproxy.its.uu.se/10.1007/s13563-021-00280-5

Tidholm, P. (2021) 'För vems skull exploateras Norrland den här gången?', *Dagens Samhälle*, 3 December.

United Nations (2015) *Transforming Our World: The 2030 Agenda for Sustainable Development*, UN General Assembly Resolution A/RES/70/1 (21 October), New York: United Nations.

7

Sacrifice Zones: A Conceptual Framework for Arctic Justice Studies?

Berit Skorstad

Introduction

Increased investment in the Arctic extractive industry over recent decades has led to new challenges for both industry itself and for society, due to the new need for minerals and rising mineral prices. With political goals such as sustainable development, climate goals and green transition, as well as an increased environmental awareness in the general population, new industrial and development projects are required to legitimize these activities both environmentally and politically (Dale et al, 2018b). This applies to new initiatives in the mining industry.

Since early major investment in the mining and mineral industry in the Nordic countries' Arctic regions, especially just after the Second World War, people have become more aware of the environmental consequences of this development. At the same time, the implications of this industry have become more visible with the use of common techniques that bring mining to the surface. Today's mining is based mostly on mountain-top removal, in contrast to underground mining. This has more consequential environmental impact by altering landscapes, removing ecosystems and emitting pollutants to land, water and air. In addition, fewer mining companies have local or national ownership, and hence less local legitimacy and social licence to operate (Skorstad et al, 2018; Prno, 2013). Conflicts around Arctic mining developments are related to local environmental and social sustainability issues, and, at the same time, divide local communities with questions about development versus the protection of traditional livelihoods (Fox, 1999; Scott, 2010; Dale et al, 2018a, 2018b).

Recent studies of the environmental consequences of industrial and mining projects have introduced the concept of 'Sacrifice Zones' (SZs) to describe the negative effect on nature, communities and human health in the immediate surroundings. Over the past twenty years, the concept's impact, popularity and application in American literature regarding nuclear testing, industrial emissions, waste sites and extractive industry have laid the foundations for asking whether the concept is also relevant beyond these contexts.

For Arctic regions, the distribution of environmental goods and 'bads' are relevant as they are often the location of extractive industries producing raw materials for a global market, while the environmental impact stays local. Trainor et al (2007, pp 627–8) state that the problem is conceived broadly as environmental inequality, as 'one in which some people bear disproportionate environmental burdens of industrial by-products or otherwise have inequitable access to environmental goods and services'. The environmental 'burdens' can be seen as a necessary side effect of industrial society and capitalism, depending heavily on input resources from nature at the same time as the system creates output waste and pollution.

This chapter posits two research questions. The main question asks, *how is the concept of Sacrifice Zones traditionally used?* Secondly, *how can Sacrifice Zones contribute to the understanding of environmental justice in the Arctic?* Included in its scope is an understanding of geographical, social and economic disparities, and differences in research traditions.

Sacrifice Zones

The concept of Sacrifice Zones (SZs) emerged in the United States after the *New York Times* wrote that Department of Energy officials reportedly described nuclear laboratories at 'superfund sites' as 'National Sacrifice Zones', being too expensive to clean up (Hedges and Sacco, 2014). Later, the concept was used in social analyses by both the media and activists. Rebecca Scott defines the concept as: 'A place that is written off for environmental destruction in the name of a higher purpose, such as national interests' (Scott, 2010, p 31); that is, describing an area that is considered lost due to environmental degradation and sacrificed for a higher (economic, national security, and so on) purpose. Others, such as Chris Hedges and Joe Sacco, have a similar description: 'areas that have been offered up for exploitation in the name of profit, progress, and technological advancement' (Hedges and Sacco, 2014, p xi). According to this connection, these zones bear the costs of industrialization, from the eradication of landscapes for the extraction of raw materials to answering the need for dumping areas for the waste from mass production and consumption. The term 'Sacrifice Zone' is used in the literature on such areas, which because of their utilitarian benefit, entail accepted environmental and social costs (Lerner, 2012).

How is the concept used?

The next section will briefly review how the concept of an SZ is used and how it is commonly framed. The sample literature chosen can be regarded as the most influential studies on the phenomenon framed as an SZ, and reveal variations in how the concept is used. The sample for this purpose includes works by Hedges and Sacco (2014), Steve Lerner (2012) and Rebecka R. Scott (2010) all of which provide different examples of SZ. This chapter also contains studies by central scholars in the field, including those by Julia Fox (1999), Danielle Endres (2012) and Ryan Holifield and Mike Day (2017).

The most prominent characterization of an SZ is the seriousness of environmental impact and the depiction of the population as marginalized. For most studies, environmental degradation has a negative impact on human health (Hedges and Sacco, 2014; Scott, 2010; Lerner, 2012), but also highly damages ecosystems (Fox, 1999; Scott, 2010). Most of the studies that we consider are based on field studies in some of America's poorest and most environmentally deprived areas. They reveal areas with a large degree of degraded environment and nature, and a population with poor health, low education and a weak economy. The use of the term SZ in connection with the environmental consequences of industrialization in rural areas appears in this literature. Some of the studies or descriptions are characterized more by activism than by traditional social science analysis. The presentation is organized into five different topics based on some recurring central themes: environmental impact, inhabitant's characteristics related to power and economic inequality, interests behind the sacrifice, the distribution of goods and burdens, and activism and social movements.

Environmental impacts

The gravity of environmental effects is prominent throughout most studies using this concept. In some studies, the SZ are areas used for military (that is, nuclear testing), hazardous waste sites or extractive activities. These zones can also be 'hot spots' where the inhabitants live in the immediate vicinity of heavily polluting industry. One example of an area labelled as SZ is the coal mining region of West Virginia, US, where the landscapes are altered due mountain-top removal techniques. Fox (1999) describes the case in West Virginia:

> The extreme conditions of exploitation of the natural and human environment ..., a Dickensian character in which relations of exploitation of both human beings and the natural environment are extremely transparent despite the fact that all of this is taking place under the mantle of economic and ecological modernization. (Fox, 1999, p 169)

The environmental impact of human activity is central in most studies of this phenomenon. This impact is both related to direct consequences for nature and the area's ecology, as well as the health of the local population. Lerner connects SZ directly to environmental problems related to pollution and illustrates this as a human rights and health issue in so-called 'fenceline' communities. However, as most studies in the literature review are social science studies, the environmental impact is described mostly as a human health problem (Lerner, 2012), the devastation of landscapes (Scott, 2010; Fox, 1999) and as endangering geographical areas local inhabitants' frame as sacred (Endres, 2012).

The difference in concept use is mostly related to how one weighs social versus environmental issues. For instance, Lerner (2012) has a greater environmental focus (that is, contaminated soil and water) than Hedges and Sacco (2014) who focus more on the socio-economic features (that is, unemployment, poverty, degraded human health) of these zones. The latter regard the sacrifice zone as the whole package of environment and social decay, while Lerner considers social and health decay because of, and in relation to, the environmental deterioration in the SZ. In this sense, the concept is strongly related to environmental justice and inequality. The point is that there is a striking and close relation between the socio-economic characteristics of the people living in these areas and the environmental state of the zone. The explanation might be that environmentally damaged areas are more affordable for groups with low income and living standards, but also that areas with low status or power might become more exposed to projects with negative environmental consequences. The latter is important when using the concept of SZ.

Environmental impacts are often disputed in SZs, and the fight for evidence is important for inhabitants and activists. One framing of the concept from cases and studies enhances a seriousness regarding encroachments on nature and human welfare in SZ.

Socio-economic characteristics

The marginalized condition of the typical SZ is well illustrated by many scholars. Without a fixed definition the term frequently reflects on the health and the way of life of low-income or minority communities (Holifield and Day, 2017). Even though Holifield and Day give nuance to this characteristic of the concept, most of the literature gives this trait special attention. This is seen in Lerner, who claims in particular that SZs are often communities consisting of low-income groups and ethnic minorities. In the portrayal of the old coalfield, Hedges and Sacco give a picture of the post-industrial society with a permanent underclass (Hedges and Sacco, 2014). They present areas of high unemployment and underemployment characterized by poverty. Their narrative consists of critical descriptions of how industry and 'corporations' exploit landscapes and people, leaving both in miserable conditions. Some (Lerner, 2012; Bullard, 2011) also underline this feature of the SZ as constituting patterns of difference in relation to environmental protection, in what they call 'environmental racism' or environmental injustice.

This marginalization can also be seen in relation to the culture and economic valuation of an area. This is typically done in studies of Indigenous communities where nature phenomena also are religious or cultural symbols (Dale et al, 2018a, 2018b; Endres, 2012). Endres (2012) uses the concept in her analysis of the conflict over the use of Yucca Mountain as an area for nuclear waste. She links the conflict in the debate to different understandings of landscapes and different values of natural areas between political authorities and Indigenous peoples. She also ties the concept to sacrificing something smaller for a larger purpose, preferably quantity over quality, and believes this must also be related to the tendency to place SZs in sparsely populated areas (Endres, 2012, p 377). The value of the area as an SZ lies precisely in this, Endres claims: 'The federal government's arguments for the Yucca Mountain site assume that it is a geologic resource to be used for its utilitarian function, in this case, a sacrifice made by a small group to benefit the entire nation' (Endres, 2012, p 334). This characterization also relates to how calculated risk is correlated with the size of the population.

As we can see, most studies argue that SZs typically affect poor states or regions in the US (such as West Virginia) due to uneven development of capitalism, social dislocation and ecological devastation (Fox, 1999). Even though the origin of the injustice seen is related to marginalization, some also address the limitations of environmental regulations in these situations. Here, environmental inequality is a concept less related to social movements than that of environmental justice. Although the American literature (Pellow, 2000; Endres, 2012) relates the concept to race and justice, studies from other regions relate it more to regions with general low income and social status.

In addition to socio-economic characteristics, it is also relevant to include the socio-cultural aspects of these areas, as poverty also can reflect a groups' or an area's political power or influence. Most of the literature analysed in this chapter also describes a lack of social and cultural capital and hence the ability to gain recognition.

Power and interests?

The question of whose interest is sacrificed and for what (or who) is also central in many studies of SZs. The answer, however, is ambiguous. Holifield and Day (2017) describe the framings varying according to how they attribute the initiators and objects of sacrifice determined by whether it is voluntary or involuntary. The inquiries on to which degree 'they' are sacrificing 'us', or 'we' are sacrificing ourselves or our local landscape, are relevant in this context (Reinert, 2018; Scott, 2010). Another question is whether the sacrifice truly is aiming for some common good or whether these are hidden in private interest. These framings can also vary in how they represent the place and scale of the originators and matters of the sacrifice. Many studies show statements that frame the primary initiators of sacrifice as an external 'they', implying that residents are being intentionally sacrificed in the interests of others. However, the answer is not as straightforward. Hugo Reinert puts this question as: 'Sacrifice thus articulates a particular relation between two concepts, such that the destruction of one brings about the gain of another. It also imputes an element of calculated, agentive will to the situation: a sacrifice does not happen by accident' (Reinert, 2018, p 599).

The motive for the local promotion of an environmentally damaging activity is often seen in relation to power and culture (Suopajärvi, 2015; Scott, 2010). One study on coal mining in the Appalachians is highly relevant, linking SZs to cultural performances (Scott, 2010). This analysis shows that parts of the local population support the development even when it entails enormous encroachments on nature. Scott's analysis of the legitimation of the sacrifice lies in the understanding of stereotypical notions of the Appalachians and the inhabitants as 'Hillbillies' and 'white trash', affecting the self-understanding of the population (Scott, 2010, p 33). Key in Scott's analysis is that sacrificing their own land is the process that gives the Appalachians status. Willingness to be a national SZ is here understood because of the Appalachians' initially low status. They become culturally required to sacrifice their landscape, their heritage and health, through coal mining to achieve normative or cultural citizenship. It is not only the presence of coal, scattered settlements and poverty that paves the way for the SZ, but also the need to increase American status, which contributes to the community (Scott, 2010).

However, the question may not only be *whose* interest but *what* interest, with the analyses often critical of the conditions that come out of 'raw capitalism' (Hedges and Sacco, 2014), that is, environmental injustice and inequality, capitalism's profit maximization and working-class powerlessness (Fox, 1999) as well as poor legal and social protection of local people (Lerner, 2012). In addition, some of the analysis also provides a deeper understanding of how race, gender and cultural perceptions reinforce the processes (Scott, 2010). Endres (2012) relates interests, opposition and injustice to power, claiming 'local opposition to proposed sites often stems from environmental injustice in the processes for site selection and local participation in decision making' (Endres, 2012, p 329). The topic of power and interests are highly related to procedural and recognitional justice. Standards of procedural justice are to do with the fairness of who is allowed to participate and be included in the process (Whyte, 2011).

Distribution of benefits and burdens

SZs are strongly characterized by uneven allocation of benefits and burdens. Fox links this to the power and predisposition of goods and burdens, saying, 'It is argued that West Virginia has become an environmental sacrifice zone, providing efficient, low-sulfur coal to the centres of accumulation and consumption at the expense of its own environment and community' (p 163). Endres (2012) makes this obvious in the case of toxic waste in general and nuclear waste in particular:

> Like other toxic wastes, nuclear waste sites tend to be sited in areas with already marginalized populations that often struggle for a voice in decision making. This is true for indigenous people, particularly in Canada, Taiwan, and the USA, raising concerns about environmental racism and nuclear colonialism. (Endres, 2012, p 329)

This study highlights such issues in a case about dumping nuclear waste in an area considered sacred by Indigenous groups, that is, using concepts such as *sacred* and *sacrifice* to effectively illustrate how landscapes, places and areas can be perceived in very different ways. This factor is highly important when it comes to valuing and assessing the impact in rural areas, as Leena Suopajärvi (2015) and Scott (2010) emphasize. As the environmental issue is obvious, so the justice aspect of it also needs to be made clear.

Lerner (2012) characterizes SZ residents as 'required to make disproportionate health and economic sacrifices that more affluent people can avoid' (2012). Scott's (2010) use of the concept underlines the human–nature relationship in context as it evokes images of incurably degraded physical landscapes, places in which not just human populations but entire ecosystems have been sacrificed.

Distributional aspects are also related to environmental justice through the idea of fairness or equity related to goods and benefits (Schlosberg, 2004). Hence, the concept has a relation to moral philosophy, like justice as fairness, and justice as mutual respect (Pellow, 2000; Rawls, 1999). Distributive justice is, however, different from standards of procedural justice, having to do with the fairness of who gets to participate, and to what degree, in the decision-making processes used to allocate risks and goods (Whyte, 2011).

Environmental inequality has emerged more recently to encompass both additional factors associated with disproportionate environmental impacts such as class, gender, immigration status, as well as the inter-connections between these factors (Sze and London, 2008). The distributional paradigm

(Schlosberg, 2004) represents not the only articulation of justice but also describes studies of environmental inequalities. This is emphasized in the inequitable share of environmental ills that poor communities, Indigenous communities and communities of colour live with. Here, the call for 'environmental justice' is relevant regarding how the distribution of environmental risks mirrors the inequity in socio-economic and cultural status. This is further related to another aspect of justice, namely justice as recognition.

Activism

The literature on SZ has a dual relationship with political activism. Some of it, like the stories by Hedges and Sacco, form part of the activism against the consequences of sacrificing communities and nature. Fox's (1999) and Learner's (2012) case studies are also investigations of environmental activism. Lerner (2012) assesses various strategies used by affected communities to improve the quality of life of citizens through corporate accountability and the government's ability to limit licensing permits. In addition, Lerner shows that strong environmental organizations can mobilize local people and reveals how lawyers can block permits or the expansion of polluting facilities, and force clean-ups of pollution. The environmental and social science research must also be seen in relation to the American social science tradition on critical theory and activism (Holifield and Day, 2017; Schlosberg, 2004).

The relation to activism also shows that the 'diagnosis' is a part of the activism, like in medicine when getting a diagnosis also brings about attention and rights. Holifield and Day (2017) suggest this discourse has helped animate mobilizations, slowing down environmental damaging projects. The framing of places and landscapes as SZ is important for building an understanding of how the SZ discourse resonates in so many different places and situations. This is relevant to residents such as those in West Virginia, where a major campaign was organized to contest mountain-top removal. This case has relevance to the Arctic, which is rich in raw materials: 'Similar to other environmental justice movements, the residents developed an understanding of the economic and political power of the coal companies and the limits of environmental regulation' (Fox, 1999, p 179).

Protests and movements are important in SZ studies, but there are also examples of divided local communities where environmentally questioned projects are welcomed by some residents, but not others (Scott, 2010), raising questions about whose interests they serve.

In summary, an SZ is characterized by a description of the ecological, economic and social costs of industrialization, where the burden is local and the gain is on a higher level. It is at the same time a compelling narrative that has spurred social movements and activists against some of the side effects caused by excessive economic development. The theoretical foundation of the concept is framed in critical realism and related to the tradition of environmental justice (Broto and Calvet, 2020).

Analytical value in Arctic justice studies

When presenting the concept SZ to scholars of Arctic studies the reaction is often that it describes something familiar, giving a sort of resonance to their own observations and experiences. As Holifield and Day (2017) state, 'Despite its conceptual ambiguities, the term sacrifice zone has become a resonant way of framing, imagining, identifying, and classifying places for the purpose of contesting activities perceived by their opponents as destructive' (Holifield and Day, 2017, p 269).

So how do the characteristics of SZs comply with the trait of the Arctic as a field? To claim a zone as sacrificed there is often talk of extreme poverty and excessive environmental damage. There are advantages and disadvantages to framing the concept strictly in this way. The advantage is that the severity of the 'sacrifice' makes the phenomenon apparent: an SZ is not just any encroachment on nature, despite the objections of the local population, but also the disproportion of bearing the goods and burdens and contesting values.

The subject is often seen as a field in social studies, showing racial and socio-economic disparities in the distribution of pollution and environmental hazards, with the environmental and social movements pointing out the problem (Mohai and Saha, 2015). For this chapter, this analytical aspect is most important. The question is whether this concept, even though it may grasp a phenomenon, also can contribute to scientific analysis. The transferability as a relevant description of communities outside its traditional field is one indication of this.

The Arctic can be seen as a geographical region, and also be described as rich in natural resources, sparsely populated, relatively low in cultural capital (education) and geographically distant from the capital (centre) of political decisions. This also applies to the Nordic Arctic region. In this context, it may therefore be a subject for sacrificing in the sense presented here (Endres, 2012; Hedges and Sacco, 2014; Scott, 2010). From this, this chapter asks whether the concept has relevance for, and whether it may contribute to, studies of environmental and social issues in the Norwegian/Nordic Arctic.

Relevance for Nordic Arctic justice studies

The Nordic Arctic generally can be described as sparsely populated, related to primary industry and little industrialization outside the extractive industry (mining and gas extraction). Both the terms 'frontier' and 'colony' have been used to describe this region's history (Brox, 1984; Aas, 1998). This may lay the ground for using the region as an SZ. On the other hand, however, the Nordic Arctic is a part of an advanced democratic welfare system with a high quality of life. The task now is to use the SZ's attributes and evaluate the potential transferability to the Nordic Arctic.

The object is not to look for areas in the Nordic Arctic that, in hindsight, can be conceptualized as an SZ. Therefore, the environmental impact is not a topic in this discussion, as that is something for an empirical study. However, the general characteristics of the area, features like industry's ownership structure, ecology, living conditions, settlement pattern, political and economic capital, and so on, may be looked upon as triggers for sacrifice.

There are surprisingly few studies done using the concept of SZ on Nordic Arctic communities. In addition to Brigt Dale, Ingrid Bay-Larsen and Berit Skorstad's *The Will to Drill: Mining in Arctic Communities* (2018a), only Hugo Reinart's 'Notes from a projected Sacrifice Zone' (2018) is observed to use the concept in this geographical area. The latter is a study of the disputed Nussir copper mine project in Northern Norway (also studied by Dale et al (2018a)). While the first discusses the relevance of this concept and illustrates that the willingness to sacrifice is highly dependent on tradition and local history (Dale et al, 2018a), Reinart (2018) describes that this motivation is related to reward in the future. The promise of a 'future of growth, prosperity, well-being for all' (Reinart, 2018, p 614) becomes the compensation for the sacrifice of nature, environment and a traditional way of living.

Even though the Nordic Arctic area often is described as a pristine nature sparsely populated by inhabitants living by and with nature – farming, fishing and herding – one also finds industry there (Dankertsen et al, 2021). The Nordic Arctic has a long history of extractive industry, particularly mining (Dale et al, 2018a). These industries have a huge impact on their environmental surroundings, the landscape, soil, air and water. In addition, the ecosystems in the Arctic are especially vulnerable (Hovelsrud et al, 2011) both due to its harsh climate and its biodiversity.

As SZ are often illustrated by extreme cases of social conditions, it may give little analytical transferability to, for example, Nordic political conditions. The Nordic political welfare model is often described as a system with a high degree of equality and generosity (Kangas and Kvist, 2018; Hvinden, 2009). The sociologist Bjørn Hvinden (2009) uses descriptions such as egalitarian values, unity and cooperation, even income distribution, low poverty, low level of conflict, high level of education, and successful mobilization of the adult population's participation.

However, statistics on living conditions have over the years shown that citizens in the Arctic parts of Norway (Nord-Norge) have relatively lower education, poorer health and a less stable income than the overall population (SSB, 2020). Even though some of these differences have decreased over the last ten years, the overall living conditions in the Arctic parts of Norway and the rest of the Nordic countries are often described as harsh. This picture is strengthened by relocation and depopulation problems, with the region being sparsely populated and with long distances between settlements and far from national centres.

Conclusion

Even though the concept of SZ has gained ground as a useful term in the critical uncovering of negative aspects of industrial development in North America, it is not difficult to find objections to its limitations. There are at least three problems with the concept used in the understanding of conflicts around extractive industries in the Nordic Arctic. Firstly, one objection is that the concept sacrifice can be misleading or ambivalent. Who performs the act of the sacrifice; for whom is this a loss? Secondly, one can question the assumption that the sacrifice is *intended* and that the SZ is valued as an SZ. A third objection may be that the concept is not relevant outside the North American political setting it is designed to describe. For example, the Nordic highly regulated political system would, one might argue, not allow such schemes.

Following elaboration of how the concept of SZ is traditionally used, this chapter seeks to answer how it can contribute to the understanding of environmental justice in the Arctic. From this perspective, it does appear to, despite its somewhat unbalanced and biased connotation. The sacrifice is seen from the local point of view. However, the perspective that lies in the concept of an SZ does not undermine the need for the development of regions and local communities, but questions how some projects fail to adequately communicate the environmental challenges to local populations. The concept of an SZ helps to see how the participation and distribution of burdens and benefits are understood and considered. It links resources and land conflicts to power, knowledge and capital.

In Nordic countries, the concept of SZ is useful to frame the result of the burden of large extraction projects on communities and ecosystems, following top-down, national policies and the global need for resources and energy. It is a combination of environmental impact, socio-economic characteristics, interests and power, the distribution of goods and burdens, and activism and social movements that are significant to evaluating the utility of SZ for the Nordic Arctic.

Study questions

1. Elaborate on the content of the concept of sacrifice zones.
2. Discuss how different aspects of the concept can be useful in the description and analysis of challenges in Arctic communities and nature.

References

Aas, S. (1998) 'North Norway-the frontier of the north?', *Acta Borealia*, 15(1): 27–41.

Brox, O. (1984) *Nord-Norge: Fra allmenning til koloni* (Northern Norway – From Common to Colony), Oslo: Universitetsforlaget.

Bullard, R.D. (2011) 'Sacrifice zones: the front lines of toxic chemical exposure in the United States', *Environmental Health Perspectives*, 119: 6. https://doi.org/10.1289/ehp.119-a266

Castán Broto, V., and M. Sanzana Calvet (2020) 'Sacrifice zones and the construction of urban energy landscapes in Concepción, Chile', *Journal of Political Ecology*, 27(1): 279–99.

Dale, B., I.A. Bay-Larsen, and B. Skorstad (2018a) *The Will to Drill – Mining in Arctic Communities*, Polar Science Series, Cham: Springer Publishing. https://doi.org/10.1007/978-3-319-62610-9.

Dale, B., I.A. Bay-Larsen, and B. Skorstad (2018b) 'The will to drill: revisiting Arctic communities', in *The Will to Drill – Mining in Arctic Communities*, Cham: Springer, pp 213–28. doi: 10.1007/978-3-319-62610-9_11.

Dankertsen, A., E. Pettersen, and J-B Otterlei (2021) '"If we want to have a good future, we need to do something about it". Youth, security and imagined horizons in the intercultural Arctic Norway', *Acta Borealia*, 38(2): 150–69.

Endres, D. (2012) 'Sacred land or national sacrifice zone: the role of values in the Yucca Mountain participation process', *Environmental Communication: A Journal of Nature and Culture*, 6(3): 328–45.

Fox, J. (1999) 'Mountaintop removal in West Virginia: an environmental sacrifice zone', *Organization & Environment*, 12(2): 163–83.

Hedges, C., and J. Sacco (2014) *Days of Destruction, Days of Revolt*, New York: Bold Type Books.

Holifield, R., and M. Day (2017) 'A framework for a critical physical geography of "sacrifice zones": physical landscapes and discursive spaces of frac sand mining in western Wisconsin', *Geoforum*, 85: 269–79.

Hovelsrud, G.K., B. Poppel, B.Van Oort, and J.D. Reist (2011) 'Arctic societies, cultures, and peoples in a changing cryosphere', *Ambio*, 40(1): 100–10. https://doi.org/10.1007/s13280-011-0219-4.

Hvinden, B. (2009) 'Den nordiske velferdsmodellen: Likhet, trygghet og marginalisering?', *Sosiologi i dag*, 39(1): 11–36.

Kangas, O., and J. Kvist (2018) 'Nordic welfare states', in B. Greve (ed.) *Routledge Handbook of the Welfare State*, Abingdon: Routledge, 148–60.

Lerner, S. (2012) *Sacrifice Zones: The Front Lines of Toxic Chemical Exposure in the United States*, Cambridge, MA: MIT Press.

Mohai, P., and R. Saha (2015) 'Which came first, people or pollution? A review of theory and evidence from longitudinal environmental justice studies', *Environmental Research Letters*, 10(2).

Pellow, D.N. (2000) 'Environmental inequality formation: toward a theory of environmental injustice', *American Behavioral Scientist*, 43(4): 581–601.

Prno, J. (2013) 'An analysis of factors leading to the establishment of a social licence to operate in the mining industry', *Resources Policy*, 38(4): 577–90.

Rawls, J. (1999) *A Theory of Justice, Revised Edition*, Cambridge, MA: Harvard University Press.

Reinert, H. (2018) 'Notes from a projected sacrifice zone', *ACME: An International Journal for Critical Geographies*, 17(2): 597–617.

Schlosberg, D. (2004) 'Reconceiving environmental justice: global movements and political theories', *Environmental Politics*, 13(3): 517–40.

Scott, R.R. (2010) *Removing Mountains: Extracting Nature and Identity in the Appalachian Coalfields*, Minneapolis: University of Minnesota Press.

Skorstad, B., B. Dale, and I. Bay-Larsen (2018) 'Governing complexity: theories, perspectives and methodology for the study of sustainable development and mining in the arctic', in B. Dale, I.A. Bay-Larsen and B. Skorstad (eds) *The Will to Drill – Mining in Arctic Communities*. Cham: Springer, pp 13–32.

SSB Statistics Norway (2020) 'Norwegian educational statistics', [online], Available from: https://www.ssb.no/utdanning/utdanningsniva/statistikk/befolkningens-utdanningsniva [Accessed 22 November 2021].

Suopajärvi, L. (2015) 'The right to mine? Discourse analysis of social impact assessments of mining projects in Finnish Lapland in the 2000s', *Barents Studies*, 1(3): 36–54.

Sze, J., and J.K. (2008) 'Environmental justice at the crossroads', *Sociology Compass*, 2(4): 1331–54.

Trainor, S.F., F. Stuart Chapin III, H.P. Huntington, D.C. Natcher, and G. Kofinas (2007) 'Arctic climate impacts: environmental injustice in Canada and the United States', Local *Environment*, 12(6): 627–43. https://doi.org/10.1080/13549830701657414

Whyte, K.P. (2011) 'The recognition dimensions of environmental justice in Indian country', *Environmental Justice*, 4(4): 199–205.

8

Planning for Whose Benefit? Procedural (In)Justice in Norwegian Arctic Industry Projects

Ragnhild Freng Dale and Halvor Dannevig

Introduction

A growing interest in Arctic resources leads to increased pressure on local authorities to accept new industrial projects in their areas. This includes mining, petroleum, wind energy and less mature technologies like hydrogen and ammonia production. This is also seen in Northern Norway where the High North (*Nordområdene*) region has become an area of strategic interest, particularly in terms of energy and security discourses (see Jensen and Hønneland, 2011; Jensen and Kristoffersen, 2013). To date, only two petroleum projects have been realized in the region: the liquefied natural gas (LNG) project Snøhvit and the oil project Goliat, both located near the town of Hammerfest. Two more fields are in the construction and planning stages, and Barents Sea petroleum continues to be controversial – though more so nationally than regionally. Conflicts over two prospective mining projects, Nussir in west Finnmark and Biedjovagge in east Finnmark, have also marked the past decade. Construction has already started in the case of Nussir, while Biedjovagge was aborted at an early stage by the municipality to avoid damaging land used for reindeer herding. These conflicts concern both the rights and interests of the Indigenous Sámi and the distribution of burdens and benefits for all parts of the region's population. New controversies have also emerged over onshore wind power and proposals for new renewable projects that are part of the transition to low carbon energy sources, and, more recently, over the increased energy demand if the electrification of the petroleum sector takes place in the north.

While mining and petroleum projects have received a lot of interest from academia (Dale, 2019; Dannevig and Dale, 2018; Magnussen and Dale, 2018; Nygaard, 2016; Arbo and Hersoug, 2010), a thorough discussion of justice which examines how theoretical approaches can be applied to processes and outcomes of energy and mining projects in the Arctic is still lacking. Examining two recent/ongoing cases from mining and petroleum, this chapter will investigate procedural, distributive and recognition (in)justice and ask what can be improved for future planning processes. We thus examine how industry regulation produces other kinds of injustices, and point to what future regulation in the region should take into account. These cases also have wider implications across the circumpolar Arctic, as the need to ensure governance mechanisms secure Indigenous rights in potential energy and mining projects is a recurring issue across the region.

Analytical framework

Recent years have seen a proliferation of academic literature on approaches to justice with regard to energy, industry, transitions and societal change. Across these different schools of justice, three components frequently appear: *distributive justice* which refers to the goal of achieving a fair distribution of burdens and benefits; *recognition justice* which states that individuals must be fairly represented, free from threats, and have complete and equal political rights (Jenkins et al, 2016; Schlosberg, 2003); and *procedural justice* which reflects how concerned actors are included in decision-making processes and to what extent they experience the process as fair (Paavola, 2004; Hiteva and Sovacool, 2017). Our focus in this chapter is on the justice aspects of the planning process for energy and mineral projects, which involve aspects of all these components: distributive justice in assessing questions like socio-economic benefits on local, national and global levels (in the form of jobs, energy provision, taxes, and so on), as well as compensation for those negatively affected; procedural justice questions (in how the process is managed); and recognition justice with regards to whom is understood as stakeholders, rights holders, and whether their rights are adequately recognized. Misrecognition leads to injustice when actors are subject to cultural domination, non-recognition and disrespect (Fraser, 1999).

The importance of *capabilities* and *participation* as salient dimensions of justice cannot be understated (see, for instance, Schlosberg, 2007). Capabilities include the ability to adapt to changing circumstances and to influence processes that enable major industrial projects to be approved. Capabilities are therefore closely related to participation, or the degree to which stakeholders are involved in and can influence processes that affect their livelihood (Schlosberg, 2007). Public participation in itself may not be enough to outweigh imbalances in power and domination, between members of the public and the energy industry or the government (Sidortsov and Katz, 2022). This is even more salient in instances where historical injustice and minority–majority societal relations are still present, as it is in most parts of the Arctic.

Legal recognition and participation do not always lead to the outcomes that are just for those the policies are meant to protect. In this chapter, we discuss matters of distribution and procedural justice, understanding it as both collective and individual with respect to Indigenous communities (Schlosberg and Carruthers, 2010). We will supplement this with a discussion about how the procedures of planning for mining and petroleum projects fall short in ensuring justice in both the processes and outcomes in the Norwegian Arctic.

Methods

The findings in this chapter are based on fieldwork in Finnmark in the period from 2011 to 2017, as well as reviews of policy documents and news media in the years between 2017 and 2021. The empirical material for the Nussir case is also presented in Dannevig and Dale (2018), and includes document analysis of the submitted assessment programme, discharge permit application, zoning plan, hearing statements, environmental impact assessment (EIA) reports, and news media articles. For the Goliat case, the material builds on insights derived from sixteen months of ethnographic fieldwork conducted between 2015 and 2017 (Dale, 2019). The fieldwork includes participant observation with time spent in the city and on the land with Indigenous and non-Indigenous locals affected in different ways, and document analysis including grey literature and media monitoring.[1]

Context Finnmark

Finnmark is Norway's northernmost and largest county. It has close to 76,000 inhabitants, making it the largest county in Norway in terms of area and the lowest number of inhabitants. Finnmark is also part of Sápmi – the homeland of the Indigenous Sámi – which stretches across the borders of what is today's Norway, Sweden, Finland and Russia. The Sámi population in Norway have faced severe 'Norwegianization' policies for more than a century, aimed at assimilating them into the Norwegian population and abandoning their Sámi language and culture. Some of these policies ended as late as the 1960s and still influence society today (Minde, 2003). The Alta

---

[1] Some of the material has been collected in collaboration with the researchers in the project 'Indigenous peoples and resource extraction in the Arctic: evaluating ethical guidelines', led by Árran Lule Sámi Centre.

struggle, which culminated in 1979/80, turned the tide on the rights of the Sámi. The planned damming of a river which would affect large Sámi areas led to the largest environmental mobilization in Norwegian history and the struggle called attention to rights of Norway's Indigenous population to their land, culture and traditional livelihoods. The aftermath of the Alta struggle led to the formation of the first committee on Sámi rights, new legislation and ratification of international treaties, and the formation of the Sámi Parliament in Norway (Minde, 2003).

The Nussir mine is planned for the Repparfjorden fjord. Repparfjorden previously belonged to the Kvalsund municipality but was merged with the Hammerfest municipality in 2020. In 2016, Kvalsund had 1,035 inhabitants distributed in several smaller settlements, of which Kvalsund is the largest. Sources of employment are primary industries (farming and fisheries) and the service industry. Kvalsund's population has steadily declined over the last three decades along with a reduced profitability from small-scale fishing and farming. The municipality has historically been a Sámi sea area, with inhabitants in small coastal settlements securing a livelihood from a combination of small-scale farming, sheep and cattle husbandry, fishing and foraging. The hills and mountain areas in Kvalsund are used as spring and summer pastures for Sámi reindeer herders in District 22/Fiettar and for parts of the season by District 20/Fálá (Magnussen and Dale, 2018).

The Goliat oil field is located 85 km from land. Hammerfest is the nearest large community where the offices of Vår Energi are located. Hammerfest has a population of approximately 10,000 people (11,000 after merging with Kvalsund), and is a regional centre with the West Finnmark hospital, a cultural centre and a city-like infrastructure with residential areas in clusters just outside the centre with shops and offices. Hammerfest lies on the Kvaløya/Fálá island, and has historically been an important hub for fisheries and the fish processing industry before this declined in the 1990s (Arbo and Hersoug, 2010). Today the business sector is dominated by petroleum, fish farms and other minor fishing activity. Many of those who live in Hammerfest today come from or have family in smaller towns and settlements that often were coastal Sámi before the assimilation policies. After decades of hidden Sáminess, there is now a growing visibility and acknowledgement of Sámi culture in the town.

The legal frameworks of mining and petroleum

Mining and petroleum in Norway are highly regulated industries with prescribed hearing rounds according to their respective sector laws. The goal of these laws is to reconcile affected interests, avoid unnecessary damage and mitigate negative impacts. In areas that concern the Indigenous population, specific laws apply with respect to consultation processes and avoiding damage to specific areas important for local customs. We will now outline the sectorial laws as well as other cross-sectorial laws that usually apply when new mining or petroleum projects are initiated.

Petroleum is regulated by the Petroleum Act, which guides the process from granting rights to survey and production licences, opening new areas for exploration, to decommissioning and liability for pollution damage. A tax-based regulation for the discharge of CO_2 in petroleum related activities on the continental shelf has been in operation since 1990. For oil and gas fields over 20 billion NOK and/or with significant societal impacts, the operator must have their Plan for Development and Operation (PDO) approved by the Norwegian Parliament before they can begin developing the field for production. Local municipalities have limited influence on the PDO, and several hearing instances are active in the hearing rounds, from local and regional authorities to nongovernmental organizations (NGOs). In the case of Goliat, the Sámi Parliament was also an important body to consult as the project is located in traditional Sámi territory.

Mining is regulated by the Mineral Act with regards to obtaining a licence to explore and extract minerals. However, any surface measures and installations will be subject to the Planning and Building Act (PBA). The municipalities enjoy a land-use planning monopoly in Norway, and thus they have the power to veto mining projects. The PBA mandates that any larger land-use change measures deliver a zoning plan with an Environmental Impact Assessment (EIA), which has to be approved by the locally elected municipal council. The scope of the zoning plan and the mandate for EIA and underlying investigations are to be presented in an assessment programme (AP), which also requires municipal approval. Both the AP and the subsequent zoning plan are subject to public hearings, as well as legal control by the county governor and several other governmental authorities.

Tailings and waste management are regulated by the Pollution Act, which mandates that the developer applies for a discharge permit. The application needs to include an EIA. When tailings are to be discharged into the sea, other laws and legal frameworks might also be invoked. In the Nussir case, the Repparfjorden is a designated 'national salmon fjord', which is a status anchored in the Biodiversity Act. Neither this nor the European Union Water Framework Directive, which has been adopted in Norwegian Law, has impacted governmental deliberations over Nussir, despite attempts made by environmental organizations to invocate this legal framework.

The municipal planning autonomy and veto power over mining projects provide the municipality ample opportunity to negotiate a Social Licence to Operate (SLO) before they approve an AP and zoning plan. Yet, there is reason to question whether these procedures secure justice for all parties, or if a process can be 'by the book' and nevertheless lead to unjust outcomes. The Sámediggi have never given their consent to the current Mineral Act as they do not think it adequately ensures Sámi rights within the traditional Sámi areas.

Municipalities do not have the same decision-making power over petroleum projects, which are regulated by the Energy Act and thereby controlled by the national government. Here, the municipality is granted a stakeholder position rather than be treated as a leader of the process. Municipalities hold some power with regulation of areas through the PBA, but as most of the petroleum development happens offshore, this is governed at the national level.

Planning process and participation of rights and stakeholders in the Goliat project

The history of petroleum extraction in Norway stretches back to the late 1960s, but is a much more recent development in Finnmark. The first exploration wells were drilled in the 1980s, but development was halted for nearly two decades before the Snøhvit project was approved in 2001. Operated by Equinor, Snøhvit is an LNG (liquefied natural gas) project with an onshore component located on a small island just outside of Hammerfest. The Snøhvit development was met with notable support and dissent by both the local and national populations. When it started production in 2007, petroleum represented something totally new for the west Finnmark region. As the region's fisheries had declined in the 1990s, Hammerfest in particular was in economic decline, and the construction work led to a flurry of activity including increasing the municipal income through the property tax, and creating jobs that would replace some of those lost with the fisheries decline (Eikeland et al, 2009).

The Goliat project is operated by Vår Energi (initially Eni Norge). It was approved in 2009 and started production in 2016 after long delays and cost overruns. The project planning stages mainly took place in the 2000s, with public hearings in 2007. Goliat was controversial for several reasons. Firstly, it was the first oil field in the Arctic part of the Norwegian Continental Shelf. Environmental NGOs feared this represented the first of many such projects in a vulnerable Arctic environment and would lead to increased CO_2 emissions. The NGOs did not succeed in stopping the development, but their involvement led to more stringent regulations, including a 'zero emission to sea' regulation and a regulation determining that waste must be deposited onshore.

Secondly, the Goliat field is close to the shore and to important fishing grounds. Unlike the gas operations tied to Snøhvit, oil operations risk polluting the precious marine environment. The fishermen's associations were initially sceptical but the number of active vessels and people employed in fishing are lower today than before the 1990s. The political risk of ignoring them was therefore smaller when Snøhvit and Goliat were approved (Hersoug, 2010). Furthermore, fishermen's associations were enrolled in a coexistence discourse (Hersoug, 2010) and local fishing vessels have become part of the emergency oil spill response. Fiskarlaget (the Norwegian Fishermen's Association) demanded that the equipment was paid for by the developer, and the 30–40 boats that are part of the emergency oil spill response also have an additional income and participate in exercises to practise a potential spill situation. As a result, the controversies in the Barents Sea have been reduced to a matter of specific areas and times of the year rather than the moratoriums that have occurred in the case of Lofoten (Hersoug, 2010).

Thirdly, local content has been an issue of controversy. Municipalities in west Finnmark were generally positive in the initial phases, when plans for Goliat included an onshore terminal somewhere along the west Finnmark coast. These expectations were created in 2006–8, when local jobs were a recurring theme in local media (Thygesen and Leknes, 2010). Eni Norge representatives visited all the local municipalities and listened to their demands, yet the discussions led to local expectations that were not fulfilled for other coastal municipalities (Dale, 2018). Though Eni *did* develop plans for both offshore and onshore concept solutions, the added cost of 10 billion NOK made it an unfeasible option. Eni's offshore solution caused outrage and disappointment, but a pragmatic acceptance of the situation was ultimately reached. Petro Arctic, for example, changed sides after the decision but made demands for local content, as did the then mayor of Hammerfest (Thygesen and Leknes, 2010).

Whilst local and regional actors had limited influence on the concept solution, lobbying for other ways to secure local content were more successful. Experiences from Snøhvit led to some changes in demands in Goliat's PDO, including a splitting of contracts into smaller parts so that local and regional firms could compete. This resulted in a relatively high number of contractors from Hammerfest and Alta and a strong growth for related industries in Hammerfest (Nilsen and Karlstad, 2016).

Sámi rights and interests were heard to a degree, but resource rights remain unresolved in Finnmark. According to the Reindeer Herding Act, the Sámi Parliament and impacted reindeer herding districts have a right to be consulted on all infrastructure projects that might have an effect on their livelihoods and traditional activities. In practice, this right to be consulted does not mean a right to veto, as the law only demands meaningful consultation and not that rights holders give their free, prior and informed consent.

When the first petroleum project was approved in 2001, the Sámi Parliament was still relatively new (Henriksen, 2010). As noted in a White Paper concerning petroleum development in the north, the Sámi Parliament claimed Norway has a commitment to pay a dividend of the income from petroleum to the Sámi people (Meld. St. 7, 2006–2007, p 34). The government at the time did not see it the same way; the Petroleum Act does not mention the Sámi people specifically, but sees petroleum as a resource that belongs to all Norwegian people alike. One of the leading bureaucrats at the time thought it 'unthinkable' that the Sámi people should have a special claim on the offshore resources (Dale, 2019, p 169).

The Goliat case shows that the local level had some (limited) influence over the project development. The PDO reflects a concern for local and regional activity, generating possibilities for jobs and for regional activity from the supply side. Yet the decision of the project as a whole remains the state's domain, particularly decisions with major economic costs. Furthermore, primary industries like fishing and reindeer herding are enrolled either into a coexistence discourse or need to adapt to the changed land use with (some) compensation.

Planning process and participation of rights and stakeholders in the Nussir project

Finnmark is rich in minerals, and in the hills behind the communities of Kvalsund and Repparfjorden in the Hammerfest municipality there are bountiful ores of copper. There was a brief instance of mining on one of these in the 1970s but prices plummeted and the mine was closed in 1978. Tailings from the mine were discharged into the sea. The old mines were then used as depots for dangerous waste and inert drilling cores from petroleum exploration in the Barents Sea as environmental regulation prevented explorers from discharging these into the sea. During the 2000s, copper prices increased significantly, and in 2007 a new ore, called the Nussir ore, was discovered by the mining company Nussir ASA. As with the previous mine operation, the plan was to discharge the mine tailings in the Repparfjord, but this time a bit further out in the fjord. Nussir submitted an AP for the mine to the Kvalsund municipality in 2008 and it was swiftly adopted by the municipal council without any alterations, despite vocal opposition by a minority of council members and some concerns from the municipal administration (Dannevig and Dale, 2018). The zoning plan was adopted by the municipal council in 2011 – again without any demands from the municipality. The same year, Nussir also submitted its discharge permit application with EIAs to the county governor.

There are mainly two causes of conflicts and concern over the Nussir project. The first is the impact of surface installations on access to the reindeer pastures crucial for Sámi reindeer herders; and the second is the discharge of tailings in the fjord. Fishermen and conservationists fear that dissolved particles from the tailings will pollute the marine environment, harming salmon, cod and other key species. The county governor placed an objection to the zoning plan as they deemed the impacts on reindeer pastures to be a violation of the rights of the reindeer herders. In 2014 the Ministry for Municipalities rejected the county governor's objection and approved the zoning plan with a condition that Nussir come to an agreement with the reindeer herders about compensatory measures. To date, no such agreement has been made.

Additionally, the discharge permit received complaints at the hearing and it was then forwarded to the Norwegian Environmental Agency (NEA) in Oslo for a decision. Due to several complaints about the quality of the EIA, the NEA requested additional assessments for how the tailings would impact marine ecosystems, but in 2016 the Ministry of Climate and Environment (MCE) granted the discharge permit. The final approval for the operation licence was granted by the government in 2019 and construction of the mine started in the summer of 2021. In response, nature conservationist organizations established a protest camp in Repparfjorden and engaged in several acts of civil disobedience, such as blocking the way for construction vehicles. The organizations also successfully petitioned the Hammerfest municipality on the grounds of some procedural errors on behalf of Nussir, and the construction work is, at the time of writing, halted. Also worth noting is that the German conglomerate Aurubis terminated a contract with Nussir over what they considered to be 'human rights issues' with respect to Nussir's conflict with the Sámi (NRK, 2021).

Distributive justice and social licence to operate (SLO) in Kvalsund

The historical experience tells us that, even though municipalities have the power to reject a zoning plan, it is the municipal council's approval of the assessment programme (with the mandate for the EIA) that precedes a zoning plan that informs the approval of a project. Therefore, Nussir initiated an influential and informative operation ahead of the approval of the assessment programme in 2011. They offered the municipality a cut of the annual profit from the mine for an industry development fund, and recruited local stakeholders in a 'reference group' that was taken on a tour to other fjords with tailings deposits and little local controversy (Dannevig and Dale, 2018). This way an SLO was established with influential stakeholders, and approval of the AP was provided.

The reindeer herders' concerns were, ultimately, largely ignored despite the Sámi Parliament and the reindeer herding district's refusal to accept the compensatory measures proposed by Nussir. There is also an issue of lack of democratic representation for the Sámi reindeer herders: reindeer herding is a semi-nomadic way of life, and herders are not represented on the municipal level in coastal Finnmark as they are registered in the Kautokeino and Karasjok municipalities where they have their winter pastures (Nygaard, 2016; Magnussen and Dale, 2018). It is also difficult to identify any attempts by Nussir to negotiate an SLO with the reindeer herders beyond the company's attempt to offer compensatory measures for the loss of pasture access as demanded by the government (Dannevig and Dale, 2018). This again qualifies as a lack of distributive justice for the Sámi reindeer herders.

The Nussir case shows that, after the approval of the AP, all subsequent decisions are made at the national level and local discourses are effectively ignored. The findings in the EIA have no bearing on the outcome, as seen in most other larger development projects (Tennøy, 2014). Thus procedural and distributive justice at the local level only exist in a 'window of opportunity' before approval of the AP. The formal knowledge basis that goes into the EIA also largely excludes local knowledge and local perspectives on valuations (Dannevig and Dale, 2018).

Discussion

Procedural justice is often viewed as fulfilled when stakeholders are engaged equitably, full and impartial information is given, and non-discriminatory procedures are followed (McCauley et al, 2013). However, as demonstrated in the examples in this chapter, the procedures followed in the legislative frameworks, EIAs and decision-making processes for mining and petroleum projects do not ensure a just outcome for all interested parties. We have focused on cases where Indigenous peoples are involved, but most of these concerns are also valid for the rights of stakeholders in other large-scale infrastructure on industry development projects. That these issues are present in a country like Norway, where trust in the democratic process is high (Schmidthuber et al, 2021) and Indigenous rights are ratified in international conventions and national law, shows that there are lessons to be learnt across Arctic states with regards to both capabilities, participation and recognition justice for Indigenous and non-Indigenous local communities alike.

In the case of Nussir, injustice experienced by the Sámi reindeer herders is first and foremost tied to misrecognition (Fraser, 1999). Procedural justice is apparently secured by the decision-making power of the locally elected municipal council as well as the hearing institute and legal protection of different rights. Recognition justice falls short in two instances: (1) the EIA does not incorporate Indigenous or traditional knowledge as it is entirely made up of external experts and consultants – this again contributes to a lack of recognition of how the Sámi value landscape qualities and their perspectives on land use; and (2) Sámi reindeer herders lack democratic representation in the municipal council and thus lose a formal venue of decision making. A new law on the duty to consult Sámi stakeholders at also the local and regional level could help to address this injustice in relation to planning and development of large-scale industrial projects.[2] Finally, the reindeer herders are also facing distributional injustice due to the unbalanced sharing of burdens and benefits associated with the mine development. The reindeer herders face a burden that threatens their livelihood but are not receiving any benefits outside the compensatory measures offered to ease access to the pastures.

The case of Goliat is more complicated as the petroleum activities are primarily located offshore. The onshore impacts are understood as indirect effects, including increased industrial activity, new power grids and the risk of damage from oil spills. Locally there is only a small chance to influence the project and local actors exercise no real power over determining the location of operations. In terms of distributive justice, the benefits mainly flow to one municipality, whereas the risk of an oil spill would disproportionally affect all the nearby coastal communities. The question of what is an acceptable risk is dependent on the incident not occurring, but is partially mitigated through the incorporation of fishing vessels in oil spill emergency preparedness.

Fishers' organizations and reindeer herders are a small proportion of the population and thus the power imbalance is marked, leading to the risk that they are overrun by more powerful actors (see, for instance, Sidortsov and Katz, 2022). Their capabilities to participate are not on equitable grounds; though the Indigenous minority is consulted, this consultation is not a right to say 'no', and the sheer size and perceived socio-economic benefit of such a project makes the impact on reindeer herding a 'minor' issue that concerns the location of specific infrastructure and timing of helicopter transport. Historical injustices continue to shape the region and its capabilities, particularly as the number of projects that will impact current land use are growing both in Finnmark and other parts of Sápmi.

Future offshore developments, and the Hammerfest LNG plant, are set to be compatible with climate regulations and emission reduction by 2030 and 2050. This implies large-scale electrification with new power grids and energy production in Finnmark, which again means the onshore impacts of offshore development will become increasingly significant. Issues of procedural justice and distributive justice should be given more attention in the decision-making process around such projects.

Conclusion

Resource extraction and industrial development have major impacts on small municipalities in the Arctic, and in this case on Sámi and coastal

---

[2] See https://lovdata.no/dokument/LTI/lov/2021-06-11-76

communities in particular. It is therefore particularly salient to understand dimensions of (in)justice of such developments, and how this is often tied to misrecognition. It also carries relevance to other cases and regions of the Arctic where reindeer herding and other Indigenous livelihoods are faced with land-use change caused by infrastructure development. The cases presented in this chapter:

- show that the ability to influence varies greatly depending on type of project and sectorial laws;
- demonstrate that it is not always clear to municipalities and impacted groups which parts of the process they can influence and how and when it is possible to do so; and
- raise issues of procedural, recognition and distributive justice, particularly for the Indigenous minority that still engages in traditional livelihoods that depend on land and sea.

The discussion in this chapter also carries relevance for other sectors, not just mining and petroleum. A recent verdict in the Supreme Court concerning wind power, the Fosen case, judged the licences of the Roan and Storheia wind power plants invalid as they hinder the south Sámi reindeer herders right to enjoy their culture.[3] This verdict shows that there are issues with the procedures of licensing for large-scale industrial projects in Norway, particularly in Sámi areas. In the future, municipalities will be given more power over the licensing process.[4]

As seen in the Nussir mining case, the municipal power to approve land-use change ensures procedural justice, underscoring that Norway's democratic institutions are well developed and highly trusted (Schmidthuber et al, 2021). This nuances insights from other studies which highlight how Indigenous groups in the Arctic are subject to procedural marginalization (Shaw, 2017); even in cases where procedural justice is fulfilled, this alone may not be enough for the outcome to be just for all parties and rights holders. Distributive injustices and misrecognition can still take place in a context where procedural justice is fulfilled, which shows an acute need to develop new means of securing recognition and distributive justice in projects that change the conditions for land use in Arctic regions. With increased interest in Arctic resources, including wind power, minerals and petroleum, such findings contain important lessons both for Norway specifically and across all Arctic regions with Indigenous populations.

---

[3] Supreme Court judgment, 11 October 2021, HR-2021-1975-S.
[4] Meld. St. 28, 2019–2020.

Study questions

1. Why is reindeer herding protected by an international treaty (the ILO 169 convention)?
2. In what ways are the Sámi reindeer herders treated by misrecognition?
3. How can the rights and interests of different stakeholders be understood from perspectives of procedural, distributive and recognition justice in cases like those discussed in the chapter?

Acknowledgements

This chapter has received funding from the European Union's Horizon 2020 research and innovation programme under grant agreement No 869327.

References

Arbo, P., and B. Hersoug (2010) *Oljevirksomhetens Inntog i Nord*, Oslo: Gyldendal akademisk.

Dale, R.F. (2018) 'Petroleum in the Barents region: local impacts and dreams at sea', in A. Szolucha (ed.) *Energy, Resource Extraction and Society: Impacts and Contested Futures*, Abingdon: Routledge, pp 37–52. https://doi.org/10.4324/9781351213943.

Dale, R.F. (2019) 'Making resource futures: petroleum and performance by the Norwegian Barents Sea', PhD Thesis, University of Cambridge, Cambridge.

Dannevig, H., and B. Dale (2018) 'The Nussir case and the battle for legitimacy: scientific assessments, defining power and political contestation', in B. Dale and I. Bay-Larsen (eds) *The Will to Drill – Mining in Arctic Communities*. Cham: Springer, pp 151–74. https://doi.org/10.1007/978-3-319-62610-9_8.

Eikeland, S., S. Karlstad, C. Ness, T. Nilsen, and I. Nilssen (2009) *Dette Er Snøhvit. Sluttrapport Fra Følgeforsknigen 2002–2008*. Rapport 2009:3. Alta: Norut.

Fraser, N. (1999) 'Social justice in the age of identity politics', in G. Henderson (ed.), *Geographical Thought: A Praxis Perspective*, London: Taylor and Francis, pp 56–89.

Henriksen, T. (2010) 'Den som tier samtykker? Samiske erfaringer fra Snøhvit', MA Thesis, University of Tromsø.

Hersoug, B. (2010) 'Fisk Og/Eller Olje?', in P. Arbo and B. Hersoug (eds) *Oljevirksomhetens Inntog i Nord*, Oslo: Gyldendal akademisk.

Hiteva, R., and B. Sovacool (2017) 'Harnessing social innovation for energy justice: a business model perspective', *Energy Policy*, 107: 631–9.

Jensen, L.C., and B. Kristoffersen (2013) 'Nord-Norge som ressursprovins: storpolitikk, risiko og virkelighetskamp', in S. Jentoft, K.a. Røvik and J.-I. Nergård (eds) *Hvor Går Nord-Norge? Politiske Tidslinjer*, Stamsund: Orkana Forlag, pp 67–80.

Jenkins, K., D. McCauley, R. Heffron, H. Stephan, and R. Rehner (2016) 'Energy justice: a conceptual review', *Energy Research and Social Science*, 11: 174–82.

McCauley, D.A., R.J. Heffron, H. Stephan, and K. Jenkins (2013) 'Advancing energy justice: the triumvirate of tenets', *International Energy Law Review*, 32(3): 107–10.

Magnussen, T., and B. Dale (2018) 'The municipal no to mining: the case concerning the reopening of the Biedjovaggi gold mine in Guovdageainnu Municipality, Norway', in B. Dale, I. Bay-Larsen and B.Skorstad (eds) *The Will to Drill – Mining in Arctic Communities*. Cham: Springer, pp 175–95. https://doi.org/10.1007/978-3-319-62610-9_9.

Meld. St. 7 (2006–2007) *Om Sametingets virksomhet*. Oslo: Ministry of Labour and Social Inclusion.

Meld. St. 28 (2019–2020) *Vindkraft på land – Endringer i konsesjonsbehandlingen*, Oslo: Ministry of Petroleum and Energy.

Minde, H. (2003) 'Assimilation of the Sami – implementation and consequences', *Acta Borealia*, 20(2): 121–46. https://doi.org/10.1080/08003830310002877.

Nilsen, T., and S.H. Stig Halgeir Karlstad (2016) 'Regionale ringvirkninger av Goliatprosjektet og Eni Norges virksomhet i Nord-Norge. Leveranser og sysselsetting i Nord-Norge av utbyggingsfase Goliat. (4/2016)', Tromsø.

NRK (2021) 'Gruveselskap mister milliardkontrakt', *Norwegian State Broadcasting*, Oslo, [online], Available from: https://www.nrk.no/tromsogfinnmark/gruveselskapet-nussir-mister-milliardkontrakt-1.15625150 [Accessed 13 May 2022].

Nygaard, V. (2016) 'Do indigenous interests have a say in planning of new mining projects? Experiences from Finnmark, Norway', *Extractive Industries Society*, 3: 17–24. https://doi.org/10.1016/J.EXIS.2015.11.009.

Paavola, J. (2004) 'Protected areas governance and justice: theory and the European Union's habitats directive', *Environmental Science*, 1: 59–77. https://doi.org/10.1076/evms.1.1.59.23763.

Schlosberg, D. (2007) *Defining Environmental Justice*, Oxford: Oxford University Press.

Schlosberg, D. (2003) 'The justice of environmental justice: reconciling equity, recognition, and participation in a political movement', in A. Light and B. Russo (eds), *Moral and Political Reasoning in Environmental Practice*, Cambridge, MA: MIT Press, pp 77–106.

Schlosberg, D., and D. Carruthers (2010) 'Indigenous struggles, environmental justice, and community capabilities', *Global Environmental Politics*, 10(4): 12–35. https://doi.org/10.1162/GLEP_a_00029.

Schmidthuber, L., A. Ingrams, and D. Hilgers (2021) 'Government openness and public trust: the mediating role of democratic capacity', *Public Administration Review*, 81(1): 91–109.

Shaw, A. (2017) 'Environmental justice for a changing Arctic and its original peoples', in R. Holifield, J. Chakraborty and G. Walker (eds) *The Routledge Handbook of Environmental Justice*, Abingdon: Routledge, pp 504–14.

Sidortsov, R., and C. Katz (2022) 'Procedural justice, affirmative and prohibitive principles', in S. Bouzarovski, T. Reams and S. Fuller (eds) *Edward Elgar Handbook on Energy Justice*, Cheltenham: Edward Elgar.

Tennøy, A. (2014) 'Kvalitet i konsekvensutredninger', in F. Holth and N.K. Wing (eds), *Konsekvensutredninger. Rettsregeler, praksis og samfunnsvirkninger*, Oslo: Universitetsforlaget, pp 185–207.

Thygesen, J., and E. Leknes (2010) 'Kampen Om Goliat. En Casestudie Av Avisenes Og Politikernes Vinklinger Av Goliat-Saken', in P. Arbo and B. Hersoug (eds) *Oljevirksomhetens Inntog i Nord*, Oslo: Gyldendal akademisk, pp 195–216.

9

The Complex Relationship between Forest Sámi and the Finnish State

Tanja Joona and Juha Joona

Introduction

Throughout the centuries, people have been keen to find and discover new areas and exploit the natural resources associated with them. According to Margaret Kohn and Kavita Reddy (2017), the term colonialism describes the process of European settlement and political control over the rest of the world, including the Americas, Australia and parts of Africa and Asia. Colonial practices were also pursued by Asian powers (Japan and Korea). The reign of European colonialism reached its apex in the 17th and 18th centuries and it began to lose relevance in the 19th century while reaching a type of endpoint in the mid-20th century. Beyond settlement practices, according to Osterhammel (2005, p 16):

> Colonialism is a relationship between an indigenous (or forcibly imported) majority and a minority of foreign invaders. The fundamental decisions affecting the lives of the colonized people are made and implemented by the colonial rulers in pursuit of interests that are often defined in a distant metropolis. Rejecting cultural compromises with the colonized population, the colonizers are convinced of their own superiority and their ordained mandate to rule.

In this, colonialism involves asymmetries in decision making and in cultural domination.

However, the imperial interests and competition between Denmark, Sweden and Russia did not focus on the overseas regions, but predominately in the areas of northern Fennoscandia known as Sápmi. The area in question was originally inhabited only by the region's Indigenous people, the Sámi (formerly known in imperial language as the Lapps). Before colonialism by the Scandinavian powers, the Sámi had their own territory. They differed from neighbouring peoples in appearance, language, clothing, culture and other similar factors. Where the neighbouring peoples earned their primary livelihood from agriculture and animal husbandry, the Sámi earned their livings from hunting, fishing and reindeer herding. Denmark/Norway, Sweden and Russia sought to extend their areas of sovereignty across these regions initially inhabited only by the Sámi and assimilate them into each of their respective empires, frequently with overlapping claims.

This development can be considered to have started as early as the 16th century (see Olofsson, 2018) as these empires expanded their territorial control. Consequently, over the centuries these states divided the territory that belonged to the Sámi. Even before the actual colonization, Sweden and Russia agreed in 1595 on an interstate border that extended north all the way to the Arctic Ocean, dividing the Indigenous group through national borders. This division continued with the boundary between Sweden and Norway, which was agreed in 1751, and the Swedish territory in the north that was divided into present-day Finland and Sweden in 1809. While frequently ignored in studies of European imperialism, the situation was in many ways, similar to the conditions of colonialism elsewhere.

The history of economic development in northern Finland has its roots in the 17th-century Settlement Bill of Lapland by King Charles IX in 1673. The Sámi were still considered landowners north of the Lapland border in the 18th century. The view was based, among other things, on the ordinances of two Swedish kings, Charles IX and John III. The settlement of Kemi and Tornio Laplands, as well as the exploitation of natural resources in a more effective manner, was the beginning of colonization in this area. This resulted in the Sámi gradually losing their ownership of land and water rights (Joona, 2019; Korpijaakko, 1989) and the resources in these places.

In a fast forward to current decision making for northern Fennoscandia, contemporary debates over this region have been fuelled by the sharp increase of interest in the use of natural resources in the Arctic, as the EU is moving towards green, ecologically sustainable forms of energy. The EU aims to be carbon neutral by 2050, and on the road to carbon neutrality the EU is committed to reducing greenhouse gas emissions by at least 55 per cent from 1990 levels by 2030. Additionally, the goal of the government programme (Government Programme of Finland, 2019) is for Finland to be carbon neutral by 2035 and the first fossil-free welfare society. The resources found in Sápmi in the Scandinavian Arctic are now at the centre of the EU green agenda and economic development.

All of the climate goals in these EU commitments are important regionally and globally in combating global warming. However, it is difficult to reconcile industries such as mining, wind power, tourism and traditional land uses (reindeer herding, hunting, fishing) in the Arctic within the green transition framework. According to Dorothee Cambou (2020), Indigenous peoples generally call for a just transition that would emphasize the recognition of their specific status and human rights. This call goes hand in hand with the understanding that respect for human rights is a key condition for sustainable and inclusive development and a means of opposing social injustices. This approach also raises questions regarding procedural and distributional justice: who bears the risks, disadvantages and benefits of economic development related to a just green transition?

In Finland, the situation is challenged by the fact that in the northernmost part of the county of Lapland – the official Sámi homeland area – industrial land use is virtually impossible. This has been established by political decisions made by the municipalities, regional government and Metsähallitus[1] (Metsähallitus, 2022). Owing to this, the pressure of land use has become even more concentrated in the areas south of the homeland area – specifically the area long inhabited by the Forest Sámi, where they hunt, fish, forage and practise reindeer husbandry up to this day (Antikainen et al, 2019).

The data used for this chapter consist of historical literature and legislation and presents an empirical case related to mining activity in the Forest-Lapland region. There is little contemporary literature on Forest Sámi, which also makes this chapter socially important and topical in connection with the EU's Green Deal (which is a strategy for a just and green energy transition).

To better understand the current situation regarding land-use pressure and Indigenous peoples' rights we will particularly focus on the *recognition, spatial, distributional* and *procedural* dimensions of justice and how they are applied in the Finnish Arctic. The use of natural resources affects the interests of different population groups in different ways. Who benefits from the use of natural resources? Who is suffering? Who should take care of environmental obligations? Whose rights are taken into account in decision making? These issues can also be seen in the context of environmental justice (see Dobson, 1998; Lehtinen and Rannikko, 2003). The debate on environmental justice criticizes the view that decision-making situations should seek to maximize the overall benefits of society (Rannikko, 2009).

Based on these conceptual definitions we ask the following research questions:

---

[1] Metsähallitus is a state-owned enterprise that produces environmental services for a diverse customer base, ranging from private individuals to major companies. Metsähallitus uses, manages and protects state-owned land and areas of water, and reconciles the different goals of owners, customers and other stakeholders. See https://www.metsa.fi/en/about-us/.

1. In terms of recognitional justice, how have the rights of the original inhabitants of Lapland to land and waters been recognized?
2. As it is difficult for the state to recognize the territorial issues related to Sámi lands in Finland, facilities or activities that cause harm to communities are currently unevenly distributed. Why do some communities suffer the effects to a significantly greater extent than others? We call this spatial or geographical discrimination.
3. What kind of procedural challenges does the current situation pose?

The following parts of the chapter contribute a historical overview of the rights and position of the Forest Sámi in Finland, and a discussion on theories and perspectives related to justice today through the lens of a case study.

A historical overview of the rights of Forest Sámi in Finland

Except for the easternmost region, which is now part of Russia, the original Sámi region is now divided into Norway, Sweden and Finland. These three Nordic countries are known as states that respect human rights, equality and the rule of law. The legal systems of the three Nordic countries are based on the same principles and have much in common, a feature especially true of Finland and Sweden, which were politically joined as the same state until 1809. In the Treaty of Hamina, which established peace between Sweden and Russia that same year, it was agreed that legislation under Swedish rule would continue to be complied with in Finland in the future, and legislation inherited from the Swedish tradition has remained the backbone of the Finnish legal system into modern times, particularly true in the context of property and real estate law – the very areas in which the land and water rights of persons belonging to the Sámi people can be considered to be based on today.

In addition to the national legal order, reference may also be made to international law. In this context, the United Nations (UN) International Covenant on Civil and Political Rights (ICCPR), (UN, 1966) and the International Convention on the Elimination of All Forms of Racial Discrimination (ICERD) (UN, 1965) should be mentioned, as Finland, Sweden and Norway have ratified them and are bound by the human rights enshrined therein, in particular with regard to the protection of Indigenous peoples and minorities. On the other hand, Finland and Sweden have not ratified the International Labour Organization's (ILO) Convention on the Rights of Indigenous Peoples No. 169 (ILO, 1989), but Norway was one of the first countries to ratify it in 1991. In Finland and Sweden, the difficulties of ratification are precisely related to the land and water rights of the Sámi people and the difficulty of recognizing them. Yet it should be asked: Does the current Finish legal system adequately respect the rights of Indigenous peoples?

Even today, the national legislation of both Finland and Sweden is primarily based on this shared tradition. In Sweden, the development that led to the recognition of Sámi land and water rights began in the late 19th century. This was done through reindeer husbandry legislation. The first reindeer husbandry law was enacted in 1886 and since then, several reindeer husbandry laws have been enacted. Today, the right of the Sámi to herd reindeer is considered a *strongly protected land use right* (Bengtsson, 2004; Allard, 2006). The right to reindeer husbandry includes the right to graze reindeer regardless of who owns the land and other land use rights such as the right to hunt and fish.

Beyond the national regulatory framework, the land and water status of the Sámi has been determined through case law. In Sweden, several dozen court decisions have shaped the special status of the Sámi in land law in its current content. It has been determined in case law that the Sámi reindeer husbandry right is a special right based on civil law. This right falls within the scope of the protection of property and is protected by the government in the same way as the right to property. In essence, Swedish legislation and case law have sought to secure the Sámi land and water rights that they have had in the past. However, in Finland, the situation is entirely different as the Finnish state considers that the Sámi do not have any special usage rights to land and water. In Finland, too, reindeer husbandry can be carried out in a reindeer husbandry area, regardless of the ownership and management of the land. In addition, this right applies to all those living in the reindeer husbandry area, not just to the Sámi.

For the above reason, no area has been defined in Finland in which the Sámi have special rights to use the land. Moreover, Finnish legislation does not specify who may have such rights but it also does not state that Sámi cannot have such rights. Therefore, there is no legal precedence set for Sámi land rights. This could be solved through legal disputes in the Finnish court system but no such action has been pursued. It is essential to understand that the Finnish state has actively avoided these discussions. No legal investigations have been carried out by, or on the initiative of, the Finnish state that seek to explain whether the Sámi have special rights to use land and water areas. The Finnish state has not even defined what land is traditional Sámi territory. Sweden, conversely, provided the Sámi these rights over a hundred years ago. In Sweden, these rights are considered to be based on national law, which is, therefore, similar to Finland.

Differences between Forest and Mountain Sámi

The area that currently forms the northernmost regions of Sweden and Finland was divided into two areas in the 17th and 18th centuries: in the coastal regions of the Gulf of Bothnia lived Finnish and Swedish peasants and

Figure 9.1: Lapp villages in northern Fennoscandia

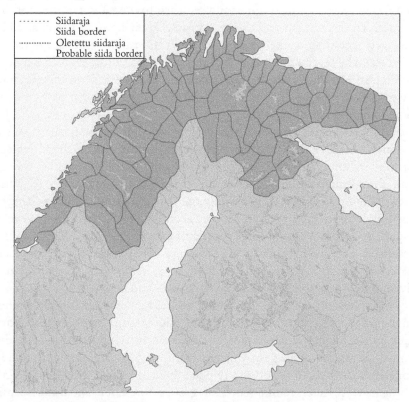

Source: Map by Johanna Roto (2005); reproduced with permission

north of that area lived the Sámi. The area inhabited by the Sámi was called Lapland, later known as historical Lapland. Already in the oldest surviving tax lists from Piitime and Luleå in 1553–1620, the Sámi are divided into two groups, that is, forest and mountain Lapps (Manker, 1968).

In Lapland, the oldest known regional division is the division into Lapp villages (*siidas*), as shown in Figure 9.1. The Lapp villages were the Sámi's own institutions. Thirteen Lapp villages were located in the whole, or in part, of the territory of present-day Finland. In that region, most of the Lapp villages were villages of the Forest Sámi. These were Inari, Kittilä, Sodankylä, Kuolajärvi, Sompio, Keminkylä, Maanselkä and Kitka. The Mountain Sámi were living in the modern-day Utsjoki area and in the mountainous area of Enontekiö.

In addition to the residential area, another factor that distinguishes these Sámi groups is their livelihoods. The livelihood of the Mountain Sámi was primarily based on reindeer husbandry, with hunting and fishing as a secondary livelihood. The lifestyle of the Mountain Sámi was also more mobile. The Mountain Sámi moved with their reindeer to the shores of the Arctic Ocean in the summer and returned to the mountain area for the winter. The Mountain Sámi were nomads. Their housing site constantly changed according to the needs of reindeer husbandry. Annual migration trips could be several hundreds of kilometres. Today, more than 75 per cent of Finnish Sámi live outside their home region, especially in the Helsinki metropolitan area, and 85 per cent of Sámi children are born outside their home region (Sámediggi, nd).

Reindeer herding was also practised by the Forest Sámi, but it was a secondary livelihood and the region had fewer reindeer than in the Mountain Sámi's territory. The focus of the Forest Sámi's livelihood was on fishing and (deer) hunting. Unlike the Mountain Sámi, the Forest Sámi lived in the area of their own Lapp village throughout the year. However, they changed their place of residence within Lapland according to a certain annual cycle, especially according to the need for hunting and fishing. In addition to the residential area and livelihoods, the difference between Forest and Mountain Sámi was especially in regards to mobility. From year to year, the Forest Sámi lived in the same area and used the same lakes and areas.

The division of the Sámi into Forest and Mountain Sámi is still reflected in the Swedish Reindeer Husbandry Act. The Swedish reindeer husbandry area is divided into Sámi villages, of which 33 are Mountain Sámi villages and 10 Forest Sámi villages.[2] Current Finnish legislation does not provide for such a division in relation to reindeer husbandry, but the Finnish Sámi Parliament Act refers to different Sámi groups (forest, fell, and fishing Sámi) (see Section 3 of the Sámi Parliament Act of 17 July 1995/974).

Land rights of Forest Sámi in the 17th and 18th centuries

The district courts led by a Swedish judge began holding hearings in the area of the Lapp villages from 1639 onwards. From the case law, it can be concluded that the Sámi belonging to each Lapp village had the *exclusive right* to decide on the use of the Lapp village area. This applied to both the management of the area and the livelihoods carried out in the Lapp village area. Finnish settlers did not have the right to settle in the area of Lapp villages without the permission of the Sámi. When the Sámi brought the dispute to the district court, the court decided that the persons in question had to leave the territory of the Lapp villages and return to the

---

[2] There are also eight 'concession villages' (*koncessiosamebyar*) located in the Kalix and Tornio river valleys. In Manker's (1968) work, *Skogslapparna i Sverige*, the Sámi who lived in the area of these river valleys have also been considered Forest Sámi.

south of the Lapland border (Fellman, 1906, pp 549–53; see also Onnela, 1995, pp 158–62). This practice continued until 1673 when King Charles XI of Sweden passed the Lapland Settlement Bill. The reasons for issuing the Bill were the labour shortage of the Nasa mountain silver mine, the needs of national defence, the promotion of agriculture, and the eradication of paganism amongst the Sámi people. The settlers were promised an exemption from all taxes and fees for 15 years and also an exemption from military service (Göthe, 1929). The new law made it possible for Finnish and Swedish peasants to settle north of the Lapland border in the area of the Lapp villages. The law did not regulate the legal status of the settlers in any way, but it remained to be decided on a case-by-case basis (Joona, 2019, pp 222–3).

The Settlement Bill of 1673 did not cause a legislative change in the land status of the Sámi. Even after the law was issued, the Forest Sámi were considered landowners in the Lapp villages areas. They were considered to have a similar right to the land they used as the peasants (Nytt Juridiskt Arkiv (NJA), 1981, pp 1, 184, 196).[3] This practice continued until 1742. In 1744, the attitude to the land rights of the Sámi changed. All land and water areas in the area of the Lapp villages were now considered to belong to the crown. Although the law does not further substantiate its position, it has emerged that this state right was justified by the doctrine of the original right of ownership of the crown and the starting point was in feudal thinking. The point of view of the state was – and still is – problematic because it is a matter of doctrine that is not based on Swedish or Finnish legislation (Joona, 2019, pp 362–71).

The legal status of Finnish Forest Sámi in Finland today – rights that are not recognized

Although the Forest Sámi who today live in the area of the former Kemi Lapland are no longer considered to have ownership of the land they used after 1744, they continue to use the land for hunting, fishing and reindeer husbandry, amongst other things. Moreover, as it became increasingly difficult to make a living from hunting and fishing under the pressure of the settlement, the Forest Sámi began to increase the number of their reindeer. Today, reindeer husbandry is the main occupation for many Forest Sámi, but hunting and fishing are also carried out. In Sweden, this right – known as the right to reindeer herding – is understood as a highly protected civil right with constitutional protection in the same way as the right to property

---

[3] The so called Taxed Mountain Case (Skattefjällsmålet), which reached the Supreme Court (Högsta domstolen) in the year of 1981.

(NJA, 1981, pp 1, 233, 248, 250; see also Bengtsson, 2004, pp 79–89; Allard, 2015, pp 248–51).

Unlike in Sweden, the Finnish state refuses to consider even the earlier Lapp villages' area a traditional Sámi area. Finnish legislation only knows the 'Sámi homeland area' provided for in section 4 of the Sámi Act, which means the areas of the three northernmost municipalities of Finland (Enontekiö, Inari and Utsjoki) as well as the area of the Lapland parish in the municipality of Sodankylä. This delimitation is based on interviews conducted in the area in 1962, in which the interviewees were asked whether the person had learned Sámi as their first language and if one of the parents or at least one of the grandparents were of Sámi origin (Nickul, 1968, p 9). When a significant number of persons meeting such criteria were found in an area, the area was referred to as the Sámi homeland area. In this respect, it can, of course, be said that few people meeting this criterion were found elsewhere, as the interviews were mainly conducted only in the area in question. It is also clear that if interviews had been conducted in a wider area and earlier than in 1962, many more people would have been found south of this area as well. The area resulting from the 1962 interviews does not constitute a special area for Sámi land and water rights, nor is it a traditionally accepted Sámi area. Despite this, the 'Sámi homeland' is the area that is usually referred to when talking about the Sámi area in modern Finland. Figure 9.2 shows the official Sámi homeland area in Finland and the situation in Sweden and Norway where the whole region of historical Lapp villages is recognized.

This current situation raises *recognition, spatial, distributional* and *procedural* issues of justice in terms of the exploitation of natural resources and the Indigenous peoples' right to use land and water areas. Unlike in Sweden, in Finland the official homeland of the Sámi is seen as a smaller region, even though there are Sámi living and practising their livelihoods outside this area. Their territory, status and rights have not been recognized in the same way as in the official homeland. In this case (economic) activities causing harm to communities are unevenly distributed. Also, their Sámi status is not taken into account in official proceedings; a good example is the Sokli mining project planned for Eastern Lapland in the former Keminkylä Lapp village area (Joona, 2020). It can be said that the starting point for everything is regional injustice, because the definition of the territory is not based on legal historical facts and is incorrectly defined in current legislation.

Reference may be made, in this respect, to the decision of the Vaasa Administrative Court in 2020. In its decision, the Administrative Court states, *inter alia*:

> The Sokli mining project is not located and its effects do not occur in the Sámi homeland referred to in section 4 of the Sámi Parliament

Act or in the Skolt area referred to in section 2 of the Skolt Act.[4] The contested decision cannot therefore be regarded as unlawful on the basis of the appeals and review concerning the examination and observance of the rights of Indigenous peoples. (Decision of the Vaasa Administrative Court 5.5.2020, decision number 20/0034/3, by Hietaniemi, Väisänen, Viitasaari and Uusi-Niemi)

A similar attitude can be found in the statements of the state authorities and the manager of the state's land assets, Metsähallitus. It is outlined in a recent plan for the use of state-owned forests that no leases will be granted for land located in the Sámi homeland, though this area would allow for the construction of wind farms. Additionally, it was stated in the same context that no mining projects in the Sámi homeland will be promoted on behalf of the state.

Figure 9.2: The official Sámi Homeland area in Finland (three northernmost municipalities)

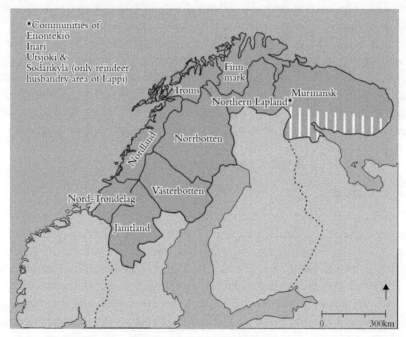

Source: Map by Johanna Roto (2005); reproduced with permission

---

[4] The Skolt-area is an area located in the municipality of Inari where the Koltta Sámi who were evacuated from the Soviet side lived after the Second World War. The Skolt area is defined in the Skolt Act (24.2.1995, p 253).

In this respect, it can be stated that the reluctance of the Finnish state to clarify and regulate the land and water rights of the Sámi has led to a very problematic situation. As late as in the middle of the 18th century, Forest Sámi in what is now Finland were considered to have the right of ownership of the areas they used, which meant a strongly protected right to the livelihoods they pursued. However, it has not been stated on behalf of the state when and how these rights lost their significance. The origin of property law thought states that rights to immovable property do not lose their significance on the sole basis of the passage of time. Yet no mention has been made on behalf of the state, which is why the immemorial prescription will not apply in the context of the rights of the Forest Sámi. In Sweden, an immemorial prescription is the legal basis on which Sámi land and water rights are considered to be primarily based. In this respect, reference may be made to the recent *Girjas* ruling of the Swedish Supreme Court (Nytt Juridiskt Arkiv (NJA), 2020).[5] The basis of this judicial institution is common to both countries, namely the real property code of the 1734 Act.

The situation would be facilitated if the traditional Sámi area were defined in legislation. This definition should be based on similar starting points as in Sweden. The current situation, where the prevailing perception is that the rights of the Sámi only apply to the so-called Sámi homeland, is incorrect in many ways. This can be said at least from a legal historical, real estate law, constitutional equality or Indigenous rights basis. It can be argued that in the areas where the Forest Sámi Lapp villages were located, more grounds can be found with the right of the Sámi to use land and water than in the Sámi mountain municipalities of Enontekiö or Utsjoki. This should be taken into account both in the preparation of legislation and in the administrative decision making of state lands.

Discussion

The historical events described in this chapter, the land use issues in the Forest-Lapland area, and the rights of Indigenous peoples to land and water are particularly challenging today because there is a constant need in the region to coordinate various land uses. Wind farm establishments and the mining industry add to the impact of ongoing state projects such as roads, forestry, hydropower and tourism activity. Together, such activities fragment the landscape and the reindeer pasture areas and create a complex impact

---

[5] The *Girjas case* is a landmark decision of the Supreme Court of Sweden. On 23 January 2020, the Supreme Court delivered its verdict (Mål nr T 853-18). See https://www.domstol.se/en/supreme-court/news-archive/a-decision-on-cancellation-of-real-estate-sales-agreements/.

pattern. Conflicting rights and conflicting issues form questions of justice and injustice.

Closely related to the just use of natural resources is the question of legitimacy and acceptability. The use of natural resources, the benefits and harms of which are perceived to be unfairly distributed, may be called into question. After a long industrial phase, Finland has been exploiting forests and nature for some time after the transition to the post-industrial period, and the relationship between wood production and other uses of nature has changed. During industrialization, most of Finland's forests were in intensive industrial use. Forestry and the forest industry provided employment and livelihoods for many professional groups across the country. As long as a large number of Finns benefited, the intensive use of natural resources was widely accepted (Rannikko, 2009).

At the moment, in connection with the green transition, there is a boom in the exploitation of natural resources in the Arctic. This situation is comparable to neo-colonialism, which, from a local perspective, appears to be an uncontrollable issue of injustice with asymmetries in decision making. Through this lens, there is a need to challenge the governance status quo and uncover social injustices in order to achieve a just transition. This includes: identifying and remedying the distributional impacts of certain development that may adversely affect certain groups (distributional justice); addressing exclusionary practices in the decision-making process that fail to include individuals and local communities in the development processes affecting them (procedural justice); and redressing the lack of recognition of pre-existing rights, needs and livelihoods of certain right-holders (recognition justice) (Cambou, 2020).

Only by gaining a broader understanding of the different theoretical approaches to justice and by looking more holistically at the challenges of land use can we build the transition to a green energy economy in a socially sustainable way. It is also clear that achieving the principles of sustainable development requires recognition and respect for human rights. People cannot be divided into different groups based on their place of residence or ethnicity, as is currently the case in Finnish Lapland. Taking all this into account is essential for the construction of the future.

Conclusion

The waves of the current debate show how challenging it is to take into account, on the one hand, legislation, the rights of Indigenous peoples and the needs of nature conservation and, on the other, the needs of people to continue to live on this planet. It is inevitable that we will have to resort to new methods of obtaining cleaner energy as climate change progresses. All of these challenges inherently raise questions about justice, as the situation of the Finnish Sámi in Finland shows, not least the distributional, procedural and recognitional aspects shaping the justice behind this development in the Finnish area of the Scandinavian Arctic.

Policy makers in municipalities are in a key position when deciding on wind farms, for example. However, they may not understand the risks involved in implementing the green transition. These are usually related to pre-existing livelihoods, traditional livelihoods, whose economic value is in many ways underestimated and which have a significant cultural value. Thus, the indirect consequences of destroying a livelihood and damaging Indigenous cultures may come as an unpleasant surprise, which future generations will criticize as a short-sighted destruction of cultural diversity. Without seeking to facilitate a just transition for local populations, it is difficult to understand the EU Green Deal as a just arrangement –despite the good intentions of the EU to fulfill its climate obligations.

The question is, have we learned anything since the 15th century? This chapter has tried to outline the challenges related to recognition, spatial, distributional and procedural justice issues from different perspectives and shows that the situation is anything but easy in Finnish Lapland. Finding a balance between Indigenous peoples' rights, green transition and economic development is a complex combination to which there are no unequivocal answers. However, it is important to study this further in the future so that we can better understand how to implement the green transition in the Arctic in a just manner.

Study questions

1. What should be the role of the state in protecting the rights of the (Forest) Sámi in Finland?
2. How and by which human rights instruments can the rights of the Sámi be protected when land use projects are planned on Sámi lands?
3. What do you think are the fundamental injustices in regard to Forest Sámi rights in Finland?

Acknowledgements
This chapter has received funding from the European Union's Horizon 2020 research and innovation programme under grant agreement No 869327.

References
Allard, C. (2006) *Two Sides of the Coin – Rights and Duties*, Lulå: Luleå University of Technology.

Allard, C. (2015) 'Some characteristic features of Scandinavian Laws and their influence on Sami matters', in C. Allard and S.F. Skogvang (eds) *Indigenous Rights in Scandinavia: Autonomous Sámi Law,* New York: Routledge.

Antikainen, J., M. Enbuske, S. Haanpää, H. Kyläniemi, V. Laasonen, M. Mayer, et al (2019) *Metsälappalainen kulttuuri ja sen edistäminen*, [online], Available from: http://urn.fi/URN:ISBN:978-952-287-763-5 [Accessed 29 March 2022].

Bengtsson, B. (2004) *Samerätt*, Stockholm: Norstedts Juridik.

Cambou, D. (2020) 'Uncovering injustices in the green transition: Sámi rights in the development of wind energy in Sweden', *Arctic Review on Law and Politics*, 11: 310–33.

Dobson, A. (1998) *Justice and the Environment*, Oxford: Oxford University Press.

Fellman, J. (1906) *Anteckningar under min vistelse i Lappmarken, Part III*, Helsinki: Rabén & Sjögren.

Göthe, G. (1929) *Om Umeå Lappmarks Svenska kolonisation från mitten av 1500-talet till omkr. 1750*, Uppsala: Almqvist & Wiksell.

Government Program of Finland (2019) Hallitusohjelma [online], Available from: https://valtioneuvosto.fi/marinin-hallitus/hallitusohjelma [Accessed 29 March 2022].

ILO (International Labour Organization) (1989) *Indigenous and Tribal Peoples Convention* (No. 169). [online], Available from: https://www.ilo.org/dyn/normlex/en/f?p=NORMLEXPUB:55:0::NO::P55_TYPE,P55_L ANG,P55_DOCUMENT,P55_NODE:REV,en,C169,/Document [Accessed 29 March 2022].

Joona, J. (2019) 'Ikimuistoinen oikeus – tutkimus Lapin alkuperäisistä maa- ja vesioikeuksista', *Juridica Lapponica*, 46, Rovaniemi: Lapin yliopisto.

Joona, J. (2020) 'One of the Finland's largest minings coming to Forest Sámi Reindeer Management Area', *Current Developments in Arctic Law*, 88, Rovaniemi: University of Lapland.

Kohn, M., and K. Reddy (2017) *Colonialism*, [online], Available from: https://plato.stanford.edu/archives/fall2017/entries/colonialism/ [Accessed 29 March 2022].

Korpijaakko, K. (1989) *Saamelaisten oikeusasemasta Ruotsi-Suomessa: oikeushistoriallinen tutkimus Länsi-Pohjan Lapin maankäyttöoloista ja -oikeuksista ennen 1700-luvun puoliväliä*, Helsinki: Lapin korkeakoulun julkaisuja.

Lehtinen, A., and P. Rannikko (eds) (2003) *Oikeudenmukaisuus ja ympäristö*, Helsinki: Gaudeamus.

Manker, E. (1968) *Skogslapparna i Sverige*, Nordiska Museet, *Acta Lapponica*, XVIII, Uppsala.

Metsähallitus (2022) [online], Available from: https://www.metsa.fi/en/about-us/.

Nickul, E. (1968) Suomen saamelaiset vuonna 1962, Tilastotieteen pro gradu-tutkielma, Helsinki.

Nytt Juridiskt Arkiv (NJA) (1981) Högsta Domstolens Dom, *Taxed Mountain Case* (Skattefjällsmålet).

Nytt Juridiskt Arkiv (NJA) (2020) Högsta Domstolens Dom, *Girjas Case* (Mål nr T 853–18).

Olofsson, F. (2018) *Discourses of Sami Rights in the Public Debate of Sweden*, Malmö: Malmö University.

Onnela, S. (1995) *Suur-Sodankylän historia 1*. Jyväskylä: University of Jyväskylä.

Osterhammel, J. (2005) *Colonialism: A Theoretical Overview*, trans. Shelley Frisch, Princeton, NJ: Markus Wiener Publishers.

Rannikko, P. (2009) 'Luonnonvarojen käytön oikeudenmukaisuus ja syrjäseudut', *Janus Sosiaalipolitiikan ja sosiaalityön tutkimuksen aikakauslehti*, 17(3): 248–57. Saatavissa, [online], Available from: https://journal.fi/janus/article/view/50525 [Accessed 29 March 2022].

Sámediggi (nd) 'Kielelliset oikeudet saamelaisten kotiseutualueen ulkopuolella', [online], Available from: https://www.samediggi.fi/kielelliset-oikeudet-ovat-perusoikeuksia/kenelle-kielelliset-oikeudet-kuuluvat/kielelliset-oikeudet-saamelaisten-kotiseutualueen-ulkopuolella/ [Accessed 26 December 2022].

UN (1965) International Convention on the Elimination of All Forms of Racial Discrimination (ICERD) Adopted by the United Nations General Assembly 21 December 1965.

UN (1966) International Covenant on Civil and Political Rights (ICCPR), Adopted by the General Assembly of the United Nations on 19 December 1966.

10

FPIC and Geoengineering in the Future of Scandinavia

Aaron M. Cooper

Introduction: why geoengineer the Arctic?

Despite the 2020 dip in carbon dioxide (CO_2), its emissions are rising as fossil fuels continue to drive the post-COVID-19 economic recovery (Friedlingstiein, 2021). This situation is being further compounded by warnings that the current climate change strategies are not being implemented at the speed required to save critical ecosystems like the Arctic (Rogeli et al, 2016). If this continues, the result could be a rise of planetary temperatures in excess of 3°C (International Institute for Sustainable Development, 2021), exceeding the 1.5°C aspirational goal of the Paris Agreement (The Paris Agreement, 2015). The Arctic forms a vital part of the cryosphere – which through surface albedo (reflectivity) is one of the ways the planet maintains its radiative balance and, thus, its temperature (Beer et al, 2020). Disruption or changes in this balance would have significant consequences on a global scale (Moon, 2021). As a result of anthropogenic global warming (AGW), the consequences of the melting Arctic icecap have become more visible (Vinnikov et al, 1999), and there are accelerated changes in the decline of the sea ice, glaciers and thawing of the permafrost (Beer et al, 2020) in a volume rate of around 3 per cent per year (Joannessen, 1999; Kashiwase et al, 2017). This decline compromises the radiative balance of the planet, resulting in increased warming and an ever-increasing risk of passing a global tipping point. Efforts at reducing both long- and short-term emissions, like black carbon and methane, are not occurring at the speed required to prevent irreversible changes (Yameinva and Kulovesi, 2018). Thus, there is a sense of urgency. The situation requires more unconventional methods like a technological intervention – geoengineering.

Research into geoengineering began to increase in the early 2000s, but it is a subject that is still relatively unknown outside academic and scientific circles – knowledge of what geoengineering precisely entails is relatively low amongst the general population – but with new developments, geoengineering is a theme that is recurrent through climate change but with some states handling novel technologies differently. The earliest adopted definition for geoengineering comes from the Royal Geographical Society. It is a 'large-scale manipulation of a specific process central to controlling the planets climate for the purpose of obtaining a specific benefit' (Royal Geographical Society, 2001). On the one hand, it has been suggested that a technological intervention like solar radiation management (SRM) through stratospheric aerosol injection (SAI) would alleviate the 'symptoms' of climate change. In turn, this would offer protection for vulnerable ecosystems like the Arctic, at least until global decarbonization can be achieved (Oxford Geoengineering Programme, 2020). But, on the other hand, there is still uncertainty over the negative effects of geoengineering and how to appropriately govern such a complex undertaking – the debate is nothing short of polarized between those in favour and those that are not. There is an increasing awareness that geoengineering the sea ice and the climate through technological intervention carries with it high risks, as the consequences will be far reaching. As Vidar Helgesen has noted: 'What happens in the Arctic, doesn't stay in the Arctic' (NATO Parliamentary Assembly, 2017).

The chapter examines elements of this polarizing debate within the context of intergenerational justice in Scandinavia. For vulnerable Indigenous populations, even though mechanisms for engagement, such as the free, prior and informed consent (FPIC) procedure exist, the implementation of geoengineering governance has to the potential to perpetuate existing colonial governance mechanisms. This effectively places Indigenous peoples in a less than adequate position. The focus lays primarily in considering questions such as: how are the costs of geoengineering to be distributed in the event of its deployment and what of the benefits? If there is deployment, how can liability be assigned in the event of an error? What are the transboundary implications if an error occurs? And, crucially, who gets to participate in the decision-making process for geoengineering projects?

Geoengineering in the Arctic and Indigenous peoples

Broadly, there are two categories of geoengineering that have emerged. The first category, greenhouse gas removal (GGR), focuses on the removal and capture of gases with high global warming potential. This can be accomplished through industrial means such as carbon capture and storage (CCS) or carbon dioxide removal (CDR), or through natural means, such as reforestation or peatland management (Global CCS Institute, 2021).

However, CCS and CDR operations are still relatively small scale – for any significant impact on the climate these would need to be scaled up (IPCC, 2021). The second category, solar radiation management (SRM) has gained a more controversial status as, although it is fraught with uncertainty, SAI has become a more serious consideration within SRM (National Academies of Sciences, Engineering and Medicine, 2021). SRM focuses on increasing the albedo of the planet, even though research into these areas has not advanced much further than laboratory modelling and simulation (NASA Earth Observatory, 2001). But on the other hand, there is still uncertainty over the negative effects of geoengineering and how to appropriately govern such a complex undertaking (Carbon Brief, 2018). There has been some testing in the Arctic where ice-geoengineering and SRM are concerned: the Arctic Ice Project (formerly Ice911) has carried out some preliminary testing in north-western Alaska, attempting to increase the surface albedo and thickness of the ice, though it has been met with an unfavourable reception.

In Fenno-Scandinavia, there has been limited engagement with geoengineering. In northern Sweden, researchers are trying potentially less invasive methods – by using a wool and corn starch blend sheet to reduce glacial melt. This has been implemented in the Kebnekaise Glacier and there have been positive results from the test (DeGeorge, 2021). But it was in February 2021 that researchers from the Keutsche Group at Harvard attempted an SAI field test with the Stratospheric Controlled Perturbation Experiment (SCoPEx) in Kiruna, Sweden (Keutsch Group at Harvard University, 2010). The Keutsche Group and its attempt to conduct an SAI experiment in Kiruna thrust the issue of geoengineering into mainstream discussion again, more specifically, in considering how SAI would physically affect the Arctic environment over the long term. This experiment was met with strong opposition from non-governmental organizations (NGOs) and the Sámi Council over the lack of consultative dialogue and the long-term physical consequences ('Open letter requesting cancellation of plans for geoengineering related test flights in Kiruna', 2021). SCoPEx received heavy criticism from the Sámi Council, which stated that such a test would lead to 'mitigation distractions' that could lead to a cascade of disastrous environmental consequences and that it should be shut down ('Support the Indigenous voices call for Harvard to shut down the SCoPEx project', 2021). Here the Sámi Council alluded to the complex atmospheric dynamics and geopolitics involved in making such an intervention, which highlights some of the deeper concerns: if geoengineering is going to be a benefit, what are the benefits and who will it benefit? The Sámi Council raised questions as to whether it was morally acceptable to conduct such a test with unclear intentions regarding the eventual deployment – especially given there was a lack of dialogue beforehand and no definitive consultation procedures, which, the Intergovernmental Panel on Climate Change (IPCC) has recommended (IPCC, 2021). This polarizing debate over the viability of governing something of this nature, like climate change mitigation, requires us to ask questions concerning the more vulnerable populations.

As engagement with these new technologies increases, there has been opposition from Indigenous peoples of the Arctic (Whyte, 2018). The Anchorage Declaration called for these 'false solutions' to be abandoned as they may be detrimental to existing participatory rights, and it further criticized the lack of affirmative action to decarbonize economies (The Anchorage Declaration, 2009; Carbon Brief, 2018; Schneider, 2022). Further, geoengineering could potentially preserve the status quo and existing power structures that have historically contributed to the subjugation of Indigenous peoples in the Arctic.

This is another part of the still polarized debate over the viability of deployment, and whether it is morally justifiable to make such an intervention given the level of uncertainty involved (McLaren and Corry, 2021), although it appears, to a degree, as though its acceptability is dependent on the level of control rather than the notion of any perceived benefits (Bellamy et al, 2017). Proponents of researching SRM have suggested that resolving these questions is not an insurmountable task, which does seem to be a valid assertion when we consider the range of mechanisms available in law (Reyolds, 2021). Further, it is thought that a more targeted application could help in maintaining the Arctic and its contribution to planetary albedo while limiting the global risks (Bodansky and Hunt (2020). There are no easy answers to these questions, as they are largely dependent on the method of geoengineering utilized, but as the Arctic becomes a focus for more concentrated efforts, we must be mindful of its impact on justice and how we can address it.

The significance of Sámi self-determination, consent and participation

So why does geoengineering pose a problem? Turning to the broader context within international law, the ability to 'consent' to any manner of relations or developments in international law is contingent on the recognition of a 'sovereign space', which is a prerequisite for the exercise of self-determination. Historically, in the crafting of sovereignty and statehood, Indigenous peoples were not granted such recognition. They were marginalized and subject to colonial rule – effectively classed as outsiders of the system (Shrinkhal, 2021). This system (and the lack of recognition of the place of Indigenous peoples) produced an inequitable distribution of social and economic benefits, which then produced injustice and claims for injustice that were seldom respected (Fraser, 2013). However, the post-Cold War Arctic saw a new world take shape. This new world would take steps to promote greater Indigenous inclusion and recognition within the region (Fitzmaurice, 2017). Generally, within the Arctic states, Indigenous peoples enjoy benefits such as: welfare, insurance, employment, recognized property rights (although not directly related to their status as Indigenous peoples) and some cultural protections. Yet within the context of the 'green transition' that is, the shift away from reliance on fossil fuels, the legacy of these colonial power structures is still evident.

Both the International Covenant on Civil and Political Rights (ICCPR) and the International Covenant on Economic, Social and Cultural Rights (ICESCR) have been key in the evolution of Indigenous self-determination within international law (Art. 1(1)): 'All peoples have the right to self-determination by virtue of the right that they freely determine their political status and freely pursue their economic, social, and cultural development.'

This did raise questions on how to adequately balance priorities. Moreover, in relation to the covenants, the United Nations Human Rights Council (UNHRC) has said that states would not be prejudiced by offering more protections under existing and future legal frameworks where Indigenous peoples were concerned – offering recognition of their place (UNHRC, 2014). In this respect, the International Labour Organization Convention No. 169 (ILO, 1989) has been a key part of this recognition as it does provide guidance on the definition of Indigenous and tribal peoples, within Art. 1(1)(b):

> Peoples in independent countries who are regarded as indigenous on account of their descent from population which inhabited the country, or a geographical region to which the country belongs, at the time of conquest or colonisation or the establishment of present state boundaries and who, irrespective of their legal status, retain some or all of their own social economic, cultural, and political institutions.

With the added issues that have arisen through delayed decarbonization – development of geoengineering as a response represents another potential avenue of colonial activity, so consent is a core part of the discussion when attempting to reconcile any potential implications. For the Indigenous peoples in the Arctic, the relationship to the environment carries cultural significance. The Sámi have knowledge of snow and ice formations and it is an integral part of their culture (Riseth et al, 2011), for example in traditional activities such as reindeer husbandry that have a unique tie to their cultural heritage and identity. Further, concepts such as common property management, the relational world view and intergenerational equity all have a significant place in Indigenous culture (Fitzmaurice, 2017), and Sámi relationships are defined by these characteristics. The United Nations Declaration on the Rights of Indigenous Peoples, UNDRIP (UN, 2007) further adds to the framework for the realization of Indigenous rights in this context. Nevertheless, the acknowledgement of these characteristics is a feature in the mandate of the Arctic Council. The Council sought to reflect these characteristics and honour these commitments when it was established in the 1996 Ottawa Declaration, noting its duty to 'Promote cooperation and interaction with the involvement of the Arctic indigenous communities'.

It is implicit then within this obligation that adequate consultation should take place with the Sámi should any development occur on what is traditionally the territory of Sápmi. Thus, based upon the existing framework, their cultural links to the environment and their ancestral land, they are entitled to being consulted before there is any preliminary testing.

Geoengineering, justice and consent

The nature of harms caused by climate change is that they are unequally distributed across the globe, from both a spatial and temporal perspective. Examining these issues through the lens of justice is still pertinent to our consideration. Whether it is through SRM or through a more direct modification of the ice, before even engaging with geoengineering, governance will need to adequately address this inability to effectively meet the requirements of justice. We can effectively relate this to the preservation of economic and political self-determination of Indigenous peoples. The FPIC in this context could be instrumental in addressing how the costs and benefits are spread when we manage the shortcomings of geoengineering.

Intergenerational justice and geoengineering the ice

SAI as a method of geoengineering is fraught with uncertainty, but a more targeted application of geoengineering may be less 'aggressive'. When it comes to geoengineering the sea ice, for example, in a manner akin to what the Arctic Ice Project is intending, there is a clearer legal framework. The basic principles of international environmental law and international human rights law form the basis of the obligations that are placed upon states. In the event of any intervention, states have the duty to ensure that both marine and human life are not adversely affected by any such activity that has the potential to cause long-term harm (UN, 1982). Further, each state has the obligation to implement appropriate environmental protection measures, including the inherent duty to consult with peoples that may be affected. From the perspective of participatory rights, human rights norms ensure Indigenous peoples have access to the necessary information so that they are aware of the risks and can informatively provide (or withhold) consent (Aarhus Convention, 2017), and this may adequately meet the requirements of intergenerational (and distributive) justice.

If we consider geoengineering in the context of the 'green transition', where a transactional paradigm has been adopted, we can draw a parallel – the idea of a cost versus a benefit. This raises concerns about the substance behind the FPIC and how we can address issues related to climate justice. With the Storheia windfarm (see Chapter 8 of this volume) Norway prioritized a move to a more renewable energy generation under the auspices of 'the green transition'. Storheia was built on what is the region of Sápmi against a background of protest which eventually led to a Norwegian high court ruling (Sámi Council, 2021; Supreme Court of Norway, 2021). The long-term benefits here are energy generation for the state and a lower level of emissions overall (more favourable for climate policies and reduction targets). But what of the cost? There is an encroachment on Indigenous land traditionally used for cultural activities like reindeer husbandry, and as a result compensation has been suggested – which is reflective of this transactional paradigm when it comes to addressing the green transition. However, while this *may* prove to be an adequate redress in the eyes of the State, it is likely to be insufficient for Indigenous groups like the Sámi because it undermines the significance of cultural activity to their overall identity. This does little to reaffirm the substance behind the FPIC in redressing the balance of intergenerational/distributive justice. Therefore, before any geoengineering projects are undertaken there must be a more robust framework that does not undermine existing protections.

Intergenerational justice and using SAI

While the more transient issues pertaining to ice-geoengineering may be easier to navigate, SAI is comparatively more complex. The justification for intervention is that achieving a more immediate result could redress climate harms using technological intervention while still thinking of the distribution of costs and benefits. There is the question of benefits and how they align with the costs. There is a huge degree of uncertainty, with evidence to suggest that intervening in one could affect another (Science Daily, 2022). The costs of climate change have already potentially compromised this future, and the use of SAI could exacerbate the situation and become equally unjust by compromising environmental quality for the future and perpetuating residual colonial power structures (Bodansky, 2020). The very notion of intergenerational equity dictates that conditions in the present do not compromise the quality of the environment for the future and the generations unborn (Brown Weiss, 2008). Even though the use of SAI does have the potential to be of benefit, questions of acceptability are prominent: where and to what extent are such interventions acceptable? What are the long- and short-term costs, and how would 'benefit' be precisely defined? If a more robust FPIC is key, the understanding of what the 'benefit' is would be crucial in any consultative dialogue. Consider the Keutsche Group and its attempted SAI experiment in Kiruna. There was a distinct lack of dialogue beforehand (Cooper, 2021).

The aim of using SAI is to bring the distribution of cost and benefits closer together, although this is not straightforward. SAI is relatively inexpensive and has the potential to effect a more rapid response with regard to atmospheric temperatures. But before it can begin to be accepted, a solution for the discrepancy between the cost and benefits needs to be distributed equally between the present and the future. This is something that is unlikely to be achieved in the current geopolitical landscape. The issues become evident when geoengineering is explored at scale. Here, there is discussion that it should be explored as a policy option. Within the geoengineering literature there are attempts to make a clear distinction between the research phase and the potential deployment (if it ever occurs) that begin to answer these questions, indicating that while research may be acceptable, deployment may not be – the lines are unclear. This is precisely why the introduction (and potential implementation) of SRM adds a further dimension to an already complicated relationship. Certain aspects of the Sámi culture are still at risk through 'green colonialism' and a 'just transition', and where consultation and FPIC are concerned, it appears that the existing legal framework is ill-equipped to deal with geoengineering, specifically SAI at this scale, especially if we are to address any potential negative consequences.

The free, prior and informed consent procedure within the context of geoengineering

Examining these questions within the context of the FPIC may nonetheless give us some direction. Considering this from the intergenerational perspective, intergenerational justice is often thought of as a form of distributional justice: the costs, harms and benefits are being dealt with over a length of time. Consider this within the context of Sen's capability approach (Jacobson and Chang, 2019), one basic requirement for an individual (or group) is that there are sufficient means for them to meet their basic requirements. To meet the requirements of intergenerational justice, geoengineering needs to have robust oversight mechanisms. If anything, it may be a catalyst for reform in numerous areas in the Arctic. Researchers should strive to develop deeper frameworks of engagement with Indigenous peoples. Although the mandate of the Arctic Council does not extend to these novel technologies, they could be included in the science-to-policy developments that are a focus of the Arctic Council. If there were a formal acknowledgement of geoengineering within the Arctic Council, it could be enough to fulfil the requirements of the FPIC (as there is a permanent voice in the form of the Indigenous permanent participants). However, there would still be concerns over how elements of justice in these circumstances would be addressed. In short, it would only solve part of the puzzle needed to resolve the intergenerational concerns. But developing its mandate to include these technologies (alongside existing legislative instruments) may facilitate the development of more robust requirements where consultation and dialogue are concerned (Smiezek, 2019). There may not be any definitive answers at this time; in fact there are many suggestions, but little by way of actualization (Corry, 2017). There will certainly need to be a robust public engagement procedure to ensure previous mistakes are not repeated.

The effects of climate change have already exposed inequities and Sweden, Finland, Norway and Russia have made commitments concerning their obligations under UNDRIP (Semb, 2012) – though there is still a reluctance to fully implement their obligations under the ILO Convention No. 169, making participation in some ways symbolic rather than substantive (Semb, 2012). Though there have been great strides in inclusion and recognition, the changes occurring due to climate change have been a catalyst for the implementation of policies and transitions that show that the colonial hierarchies still define the relationships between the state and Indigenous peoples. The FPIC has been key in mitigating some of the influence that state sovereignty possesses over Indigenous peoples – consider how the rights in Article 27 of the Covenant (ICCPR) protect the link between territory and the realization of Indigenous self-determination. The social and environmental dimensions of geoengineering will have far reaching effects beyond the atmosphere (Parker et al, 2020), so addressing the inadequacy of the current consultation steps is required. Yet, the exercise of Sámi autonomy through the rights laid out in the legal framework is often contingent on the priorities of each respective state: land-use and sustainable development initiatives under the auspices of the 'green transition' have already placed a strain on the ability of the Sámi to maintain traditional cultural activities (Sámi Council, 2021).

These links are precisely why the principle of FPIC has been vital in the exercise of self-determination. ILO, UNDRIP and the Convention on Biological Diversity (specifically through the Nagoya Protocol) have been key in establishing a formal right to consultation and cooperation. Further, where consultation is concerned, the Aarhus Convention has been vital in adding further robustness to procedure through the access to information, participation in decision making, and, crucially, the right to access justice in matters concerning the environment. But where to begin in devising a process that would adequately provide consultation? Governance implications within the context of geoengineering are uncertain and threaten to subordinate Indigenous peoples in the decision making process.

Conclusion: Can a more robust FPIC provide a solution?

> The protection of these rights is directed towards ensuring the survival and continued development of the cultural, religious and social identity of the minorities concerned, thus enriching the fabric of society as a whole.
>
> Human Rights Committee, 1994

In short, it is unlikely. This chapter has focused upon raising awareness of these complex issues. Human civilization is unique in that in most cases our presence in an environment is immediately apparent, and we can make large-scale alterations to the environment to suit our needs. Climate change and global warming is a consequence of an excessive ability to change the natural world. Overall, it is a scenario that has been created by the behaviour of a group of nations that since the industrial revolution has continued to benefit a distinct group of individuals. Now is a time where the relationship between nature and the human race is being redefined. But how this relationship evolves is often dictated by our social values and technological development. It has been shown that technology interacts with our value and belief systems; it alters behaviours – both conscious and subconscious. In this context, regardless of the whether geoengineering is an inevitability, it is not an exceptional concept that geoengineering could provide a novel technological solution to a problem.

There is a great degree of apprehension around geoengineering and how it could potentially preserve this status quo (practically and legally) – effectively leading to the perpetuation of the colonial hierarchies which have essentially laid the foundations for the situation we see with geoengineering (and its inability to cope with the requirements of justice). While we could potentially stave off the more serious consequences of climate change, we are still allowing the highest emitters of greenhouse gases to continue (Zhen et al, 2021). The green transition on the whole has been somewhat of a detriment to the participatory rights of the Sámi. Even though there is some recognition (in terms of guaranteeing economic and cultural self-determination) it is still quite limited. The Scandinavian states have not fully implemented their international obligations when it comes to the protection of Indigenous groups. In terms of success stories there is little that could be provided when it comes to engagement and benefit sharing. Communicative planning scholars often claim that forms of participatory planning centred on public deliberation can facilitate more equitable decision making by overcoming power differentials between citizens and stakeholders. The FPIC as a procedure is ineffective and its implementation rests upon the cooperation of the states involved, which is contingent upon the balance within the states. Consequently, the emphasis here is on the construction of a robust system to tackle these challenges. Intergenerational justice depends upon laws designed to hold states and corporations accountable for pollution and rights violations and their enforcement by courts willing to acknowledge public alarm about global heating. For the Arctic, when it comes to geoengineering, it must tread carefully when engaging with these tools.

Study questions

1. Given that the emergence of geoengineering could detrimentally affect vulnerable, Indigenous groups, how can FPIC within the context of geoengineering help further develop restorative justice within the portfolio broader climate change solutions?
2. What can be done to redress the issues causes by the cost/benefit paradigm within the context of geoengineering?
3. How could an intergovernmental forum akin to the Arctic Council regulate consultation and dialogue on the research and develop of geoengineering in the Arctic?
4. Why, and how, could this be an opportunity for Arctic Council reform?

References

Beer, E., I. Eisenman, and T.J.W. Wagner (2020) 'Polar amplification due to enhanced heat flux across the halocline', *Geophysical Research Letters*, 47(4). https://doi.org/10.1029/2019GL086706.

Bellamy, R., J. Lezaun, and J. Palmer (2017) 'Public perceptions of geoengineering research governance: an experimental deliberative approach', *Global Environmental Change*, 45: 194–202. https://doi.org/10.1016/j.gloenvcha.2017.06.004.

Bodansky, D., and H. Hunt (2020) 'Arctic climate interventions', *The International Journal of Marine and Coastal Law*, 35(3): 596–617. https://doi.org/10.1163/15718085-BJA10035.

Brown Weiss, E. (2008) 'Climate change, intergenerational equity, and international law', *Vermont Journal of Environmental Law*, 9: 615–27.

Carbon Brief (2018) 'Geoengineering carries "large risks" for the natural world, studies show', [online], Available from: https://www.carbonbrief.org/geoengineering-carries-large-risks-for-natural-world-studies-show/. [Accessed 20 December 2022].

Cooper, A.M. (2021) 'Sámi Council resistance to SCoPEX highlights the complex questions surrounding geoengineering and consent', *The Arctic Institute – Center for Circumpolar Security Studies*, 20 May, [online], Available from: https://www.thearcticinstitute.org/sami-council-resistance-sco pex-highlights-complex-questions-geoengineering-consent/ [Accessed 26 October 2022].

Corry, O. (2017) 'The international politics of geoengineering: the feasibility of plan B for tackling climate change', *Security Dialogue,* 48(4): 297–315. https://doi.org/10.1177/0967010617704142.

DeGeorge, K. (2021) 'A cloth sheet helped protect a Swedish glacier from global warming', *ArcticToday* [blog], [online], Available from: https://www.arctictoday.com/a-cloth-sheet-helped-protect-a-swedish-glacier-from-global-warming/. [Accessed 20 December 2022].

Fitzmaurice, M. (2017) 'Indigenous Peoples and Intergenerational Equity as an Emerging Aspect of Ethno-Cultural Diversity in International Law', in G. Pentassuglia (ed.) *Ethno-Cultural Diversity and Human Rights*, Nijhoff: Brill, pp 188–222.

Fraser, N. (2013) *Fortunes of Feminism: From State-Managed Capitalism to Neoliberal Crisis*, London: Verso Books.

Friedlingstein, P. (2021) 'Global carbon budget 2021' (Global Carbon Project 2021), [online], Available from: https://essd.copernicus.org/preprints/essd-2021-386/essd-2021-386.pdf [Accessed 4 January 2022].

Human Rights Committee (1994) General Comment 23, Article 27 (Fiftieth session, 1994), Compilation of General Comments and General Recommendations Adopted by Human Rights Treaty Bodies, U.N. Doc. HRI/GEN/1/Rev.1 at 38 (1994).

International Labour Organization (1989) Indigenous and Trible Peoples Convention No. 169.

IPCC (2021) *Climate Change 2021: The Physical Science Basis*, Contribution of Working Group I to the Sixth Assessment Report of the Intergovernmental Panel on Climate Change, edited by V. Masson-Delmotte, P. Zhai, A. Pirani, S. L. Connors, C. Péan, S. Berger et al (eds), Cambridge: Cambridge University Press, [online], Available from: https://www.ipcc.ch/sr15/ [Accessed 19 April 2019].

Jacobson, T., and Chang, L. (2019) 'Sen's capabilities approach and the measurement of communication outcomes', *Journal of Information Policy*, 9: 111–31.

Johannessen, O.M., E.V. Shalina, and M.W. Miles (1999) 'Satellite evidence for an Arctic sea ice cover in transformation', *Science*, 286: 1937–9, [online], Available from: https://www.science.org/doi/10.1126/science.286.5446.1937 [Accessed 20 December 2022].

Kashiwase, H., K.I. Ohshima, and S. Nihashi, et al (2017) 'Evidence for ice-ocean albedo feedback in the Arctic Ocean shifting to a seasonal ice zone', *Scientific Reports*, 7, [online], Available from: http://www.nature.com/articles/s41598-017-08467-z [Accessed 20 December 2022].

Keutsch Group at Harvard – SCoPEx (2010), [online], Available from: https://www.keutschgroup.com/scopex [Accessed 20 December 2022].

McLaren, D., and O. Corry (2021) 'The politics and governance of research into solar geoengineering', *WIREs Climate Change,* 12(3). https://doi.org/10.1002/wcc.707.

Moon, T.A. (2021) 'NOAA Arctic Report Card 2021 Executive Summary', [online], Available from: https://library.oarcloud.noaa.gov/noaa_docume nts.lib/OAR/GOMO/Arctic_Report_Card/ARC21_Executive_Summ ary.pdf [Accessed 10 February 2022].

NASA Earth Observatory (2021) 'Global effects of Mount Pinatubo', 15 June, [online], Available from: https://earthobservatory.nasa.gov/images/ 1510/global-effects-of-mount-pinatubo [Accessed 23 March 2022].

National Academies of Sciences, Engineering and Medicine (2021) *Reflecting Sunlight: Recommendations for Solar Geoengineering Research and Research Governance*, Washington, DC. https://doi.org/10.17226/25762.

NATO Parliamentary Assembly (2017) ' "What happens in the Arctic, does not stay in the Arctic": climate change in the Arctic will have global consequences and cannot be ignored', [online], Available from: https://www. nato-pa.int/news/what-happens-arctic-does-not-stay-arctic-climate-cha nge-arctic-will-have-global-consequences [Accessed 20 December 2022].

'Open letter requesting cancellation of plans for geoengineering related test flights in Kiruna' (Sámiráđđi) (2021), [online], Available from: https:// www.saamicouncil.net/news-archive/open-letter-requesting-cancellat ion-of-plans-for-geoengineering [Accessed 11 April 2022].

Parker, A., J.B. Horton, and D.W. Keith (2020) 'Stopping solar geoengineering through technical means: a preliminary assessment of counter-geoengineering', *Earth's Future*, 6(8): 1058–65. https://doi.org/10.1029/2018EF000864.

The Paris Agreement (2015), [online], Available from: https://unfccc.int/ sites/default/files/english_paris_agreement.pdf [Accessed 4 January 2022].

Reynolds, J.L. (2021) 'Is solar geoengineering ungovernable? A critical assessment of governance challenges identified by the Intergovernmental Panel on Climate Change', *WIREs Climate Change,* 12(2). https://doi. org/10.1002/wcc.690.

Riseth, J.Å., H. Tømmervik, E. Helander-Renvall, N. Labba, C. Johansson, E. Malnes, J.W. Bjerke, et al (2011) 'Sámi traditional ecological knowledge as a guide to science: snow, ice and reindeer pasture facing climate change'. *Polar Record*, 47(3): 202–17. https://doi.org/10.1017/S0032247410000434.

Rogelj, J., M. den Elzen, and N. Höhne, et al (2016) 'Paris Agreement climate proposals need a boost to keep warming well below 2°C', *Nature,* 534: 631–9, [online], Available from:http://www.nature.com/articles/ nature18307 [Accessed 19 December 2022].

Royal Geographical Society (2001) 'Geoengineering', [online], Available from: https://www.rgs.org/geography/online-lectures/geoengineering/ [Accessed 20 December 2022].

Sámi Council (2021) 'Sámi victory in Supreme Court – illegal wind farm on Sámi land' (Sámiráđđi), [online], Available from: https://www.saami council.net/news-archive/smi-victory-in-supreme-court-illegal-wind- farm-on-smi-land [Accessed 18 April 2022].

Schneider, L. (2022) 'High-risk geoengineering technologies won't reverse climate breakdown', *Geoengineering Monitor*, 2 March 2022, [online], Available from: https://www.geoengineeringmonitor.org/2022/03/high-risk-geoengineering-technologies-wont-reverse-climate-breakdown/ [Accessed 23 March 2022].

Science Daily (2022) 'Geoengineering could return risk of malaria for one billion people', [online], Available from: https://www.sciencedaily.com/releases/2022/04/220420092126.htm [Accessed 20 December 2022].

Semb, A.J. (2012) 'Why (not) commit? – Norway, Sweden and Finland and the ILO Convention 169', *Nordic Journal of Human Rights*, 30(2): 122–47, [online], Available from: https://www.idunn.no/doi/abs/10.18261/ISSN1891-814X-2012-02-02 [Accessed 27 May 2022].

Shrinkhal, R. (2021) '"Indigenous sovereignty" and right to self-determination in international law: a critical appraisal', *AlterNative: An International Journal of Indigenous Peoples*, 17(1), [online], Available from: http://journals.sagepub.com/doi/10.1177/1177180121994681 [Accessed 20 December 2022].

Smieszek, M. (2019) 'Evaluating institutional effectiveness: the case of the Arctic Council', *The Polar Journal*, 9: 3–26.

'Support the Indigenous voices call on Harvard to shut down the SCoPEx project' (Sámiráđđi) (2021), [online], Available from: https://www.saamicouncil.net/news-archive/support-the-indigenous-voices-call-on-harvard-to-shut-down-the-scopex-project [Accessed 11 April 2022].

Supreme Court of Norway (2021) 'Licences for wind power development on Fosen ruled invalid as the construction violates Sami reindeer herders' right to enjoy their own culture', [online], Available from: https://www.domstol.no/en/supremecourt/rulings/2021/supreme-court-civil-cases/hr-2021-1975-s/ [Accessed 20 December 2022].

The Anchorage Declaration (2009), [online], Available from: https://unfccc.int/resource/docs/2009/smsn/ngo/168.pdf [Accessed 20 December 2022].

The Aarhus Convention (2017) European Commission, [online] Available from: http://ec.europa.eu/environment/aarhus/ [Accessed 20 December 2022].

UN (1982) *Convention on the Law of the Sea*, 10 December 1982.

UN (2007) Declaration on the Rights of Indigenous Peoples : Resolution. Adopted by the General Assembly, 2 October 2007, A/RES/61/295.

UNHRC (2014) Contribution of OHCHR's Indigenous Peoples and Minorities Selection Contribution to the Thirteen session of the UN Permanent Forum on Indigenous Issues, 2014, [online], Available from: https://www.un.org/esa/socdev/unpfii/documents/2014/ohchr.pdf [Accessed 19 December 2022].

Vinnikov, K.Y., A. Robock, R.J. Stouffer, J.E. Walsh, C.L. Parkinson, D.J. Cavalieri, et al (1999) 'Global warming and northern hemisphere sea ice extent', *Science*, 286: 1934–7, [online], Available from: https://www.science.org/doi/10.1126/science.286.5446.1934 [Accessed 19 December 2022].

Whyte, K.P. (2018) 'Indigeneity in geoengineering discourses: some considerations', *Ethics, Policy & Environment,* 21(3): 289–307. https://doi.org/10.1080/21550085.2018.1562529.

Yamineva, Y., and K. Kulovesi (2018) 'Keeping the Arctic white: the legal and governance landscape for reducing short-lived climate pollutants in the Arctic region', *Transnational Environmental Law,* 7(2): 201–27. https://doi.org/10.1017/S2047102517000401.

Zhen, D., E.T. Burns, P.J. Irvine, D.H. Tingley, J. Xu, and D.W. Keith (2021) 'Elicitation of US and Chinese expert judgments show consistent views on solar geoengineering', *Humanities and Social Sciences Communications,* 8(18), [online], Available from: https://www.nature.com/articles/s41599-020-00694-6 [Accessed 17 January 2022].

11

Overarching Issues of Justice in the Arctic: Reflections from the Case of South Greenland

Joan Nymand Larsen and Jón Haukur Ingimundarson

Introduction

Arctic communities must negotiate locally, regionally and with external actors on how to make the best use of the human and natural resources in the region. This is against the backdrop of increased globalization, climate change, high demand for the region's mineral resources, and increasing local demand for improved living conditions and sustainable livelihoods.

The regional and local contexts for achieving improved living conditions and more sustainable economic development vary across the Arctic. While Arctic economies have many characteristics in common – basic structural pillars and gaps in financial, human, physical and natural capital that frequently interfere with progress in economic development – the nature of their differences is what sets them apart. Their level of internal resilience differs widely, and therefore so does their ability to respond (Larsen and Huskey, 2015). Heterogeneity means that while local economies are subject to similar economic signals and disturbances from external environments, they respond differentially to regional and global changes. Differences in capacity to respond are linked to their broad diversity in human, physical, social and natural capital (Larsen and Petrov, 2020). Some of these capitals are limiting factors and cause inequalities between different contexts.

This chapter explores some of the key issues and challenges of justice and injustice in the Arctic with a focus on the case of South Greenland.

Historical background

The history of Greenland depicts a country undergoing transition and with socio-economic changes on multiple fronts. Changes in governance, institutions and general economic structures continue to have profound and diverse implications for local communities and Indigenous livelihoods and traditions. Greenland today is a country experiencing rapid and multiple changes: environmental, economic and social. Yet these overwhelming changes began for Greenlanders in the age of colonization. The Danish colonial period in Greenland began approximately 300 years ago and throughout its early history colonies were established along the coast of Greenland. The Danish trade monopoly Kongelige Grønlandske Handel (KGH) became a central feature of the colonial period and operated from the time it was established in 1776 until the end of the Second World War. Greenland became an integral part of the Kingdom of Denmark when its colonial status was abolished in 1953. No real changes in the administrative ties between Greenland and Denmark took place with the end of the colonial period, however, as Denmark continued to administer the common civil rights and govern Greenland with the same civil servants and the same colonial administrative body (Larsen, 2002).

The first phase of economic development policy in Greenland began in 1953, with the G50 (a Danish Royal Commission report on the development of Greenland in the 1950s). The overall objective of G50 was to create greater equality between Greenland and Denmark, to improve the standard of living, and to establish a higher degree of economic independence for Greenland. The Danish administration sought to achieve this through population concentration, importing Danish capital and personnel, the privatization of state-run operations, investments in infrastructure and the modernization of the fishing industry. To achieve centralization as quickly as possible, targeted localities were often denied housing and business support while KGH discontinued its investment programme and the maintenance of production and fish processing plants. For the administration, this had the intended effect of increasing the speed of centralization. The number of settlements fell significantly over time due to the pricing and resupply policies, which exerted profound economic hardship on households and, in the period that followed, the integration of families from settlements into urban and central towns caused profound and life-altering changes for many. The negative socio-economic outcomes experienced by many from these policies, and the cultural and social challenges and inequalities that persist, are often linked to these historical events. Overall, economic development in Greenland around the mid-1960s remained grossly disappointing not only because Greenland lacked educational facilities, but also because significant amounts of value added, resource rents, profits and income generated in connection with development projects did not remain in Greenland but went to Denmark (Larsen, 2002).

In 1964 the second phase of economic development policy in Greenland, G60, was implemented and dominated the Danish development policy up until the introduction of Home Rule in 1979. The Danish state made provisions for increased activities, which included state support for the acquisition of larger trawlers and the construction of fish processing plants. These initiatives were largely a response to the disappointing private initiative in Greenland, which resulted from the high start-up and operating costs in the private sector caused by the scarcity of critical resources, the high costs of importing intermediate products, transportation to markets, and the cost of infrastructure. The role of the Danish state became first and foremost to establish the technical and financial requirements for a centralized and industrialized fishery (Larsen, 2004).

In the 1970s discontent with the Danish administration in Greenland was growing. This was fuelled in part by economic development promises that had failed following the end of the colonial period; ethnic stratification in Greenland between Greenlanders and Danes; the control of trade and commerce by the Danish trade monopoly; Greenland's objection to membership in the European Economic Community (EEC); and the almost complete control and influence by Danish authorities over the design of the Greenlandic school system, which has had far-reaching and overwhelming consequences for Greenland, including persisting language challenges and gaps in educational attainment to this day.

When Greenland was granted Home Rule in 1979 it was a step closer towards self-government, and efforts were now placed on developing a more diversified and resilient economy. While the fishing industry had proven itself as the backbone of Greenland's economy, the initiatives that followed were increasingly focused on developing alternative sources of income within the mineral resources and tourism industries as well as other land-based trades.

A key objective in the 1980s was the further modernization of Greenland's fishing industry. This included the modernization of fish processing plants in the towns and the restoration of worn-out fish processing facilities in settlements, and greater access to low interest loans for investing in trawlers and fishing boats (Poole, 1990; Danielsen et al, 1998; Larsen, 2004). When Greenland left the EEC in 1985 it obtained OLT status (overseas territory in relation to the EEC), and thereby duty-free access to EEC markets in exchange for EEC fishing rights. This addressed critical issues concerning economic inequality and the lack of autonomy with regard to this key source of income for the country.

Reforms introduced in the 1990s signified a move toward a free-market ideology. This included a reduced public sector role, industry restructuring, rationalizations, contracting out, and the start of discussions over the removal of the uniform price system which was considered a precondition for a more competitive market structure and more privatizations (Larsen, 2004). The uniform price system had all towns and settlements pay the same price for utilities no matter their cost of provision. This system worsened the economic conditions for those who were already living on the margin, thereby increasing inequalities.

The conversion of the fishing, production and export business into the Home Rule owned company, Royal Greenland Inc., in the 1990s was the first step to developing a business community operating under market conditions. The commercialization of Home Rule owned enterprises was undertaken to separate the political agenda from the economic agenda, thereby making social assignments, such as maintaining production, supply and employment in outlying districts, the responsibility of separate, non-commercial enterprises. This move affected existing power asymmetries and increased inequalities between towns and settlements.

Since obtaining self-government in 2009, the government of Greenland has looked for new sources of income – including mineral extraction – that can further reduce dependence on Danish annual block grants (Larsen, 2010).

In 2009 Greenland also underwent a significant municipality restructuring which meant a reduction in the number of municipalities from 18 to its current five. In South Greenland, three municipalities (Nanortalik, Narsaq, Qaqortoq) were amalgamated into one municipality, namely Kujalleq, with the administrative centre located in Qaqortoq. Today many Nanortalik and Narsaq residents express feelings of disempowerment and lack of control over their lives. These sentiments signal a real change in the balance of power that, according to many locals, has left them poorer compared to their relatively favourable economic conditions prior to restructuring. The sense of disempowerment and perceived lack of justice is often described in relation to feelings of being left behind and not being included in decisions that affect their lives.

The balance of power

In many parts of the Arctic competing rights and interests related to the use of renewable and non-renewable resources present a source of conflict between different stakeholders, and decisions regarding resource use and allocation produce winners and losers (Larsen, 2010; Duhaime et al, 2017; Larsen and Huskey, 2020).

While economic growth has been a key policy aim – with increased investments in infrastructure, education, economic sectors of raw minerals, tourism, fisheries and other land-based trades – such policy objectives can have undesirable societal consequences, as the depletion of scarce natural resources, environmental degradation and human costs from changed livelihoods may result. The more recent Greenland policy objectives for increasing the ability of the raw mineral sector to boost its economic significance almost resulted in the opening of a rare-earth and uranium mine in Narsaq, South Greenland, despite persistent local resistance due to fear of environmental and social impacts. In the 1950s uranium and rare-earths deposits were discovered at Kuannersuit (near Narsaq) and led to plans for uranium extraction that were later abandoned by the Danish government in 1983. In 2007 Greenland Minerals and Energy acquired the area, and due to the Greenlandic government's relaxation of regulations in 2010, a mining licence application for an open pit mine was submitted in 2015. In 2021 Greenland's parliament passed legislation that bans uranium mining, and therefore the mining project was no longer going ahead.

While resources may leave the Arctic region in vast quantities, their exploitation can give rise to positive local and regional spin-off effects related to extraction, and important indirect effects with secondary activities can develop to help support industry and small land-based trades. Benefits to economic growth associated with the primary resource trade may include improvements in the utilization of existing factors of production, increased factor endowments and economic linkage effects.

Addressing the key policy objectives of growth, efficiency, equity and stability may create conflicts of interests and rights between different Arctic stakeholders with competing interests concerning issues that may be highly emotionally charged, such as those concerning the environment and the allocation and use of natural resources (Larsen, 2010). Those who have greater leverage than others because of their larger endowments of critical resources often tilt the scales and ongoing developments can move in their favour. Having a voice and an opportunity to participate in decision making at the local level is a key determinant of the local economic outcomes and for a more equitable distribution of benefits and costs related to development that affect people's lives (Aarsæther et al, 2004; Rasmussen et al, 2014).

Economic diversification can make a place more resilient to external shocks and disturbances as it provides opportunities for moving resources between sectors. Examples of this are present in the case of Nanortalik and Narsaq, where both places have experienced periods of downswings in their fisheries, which for extensive periods has caused long-term unemployment due to a lack of alternative trades and economic opportunities. In more recent times, however, the growth of tourism (and in the case of Nanortalik, the Nalunaq gold mine only 30 km from town) has created more resilience and greater stability until the recovery in fisheries. In the case of Narsaq, agriculture, including sheep and cattle farming, is also a promising sector of growth.

Local control and ownership may be better realized by investing in smaller-scale activities such as arts and crafts, tourism, agriculture and small-scale trade locally. When these activities require start-up financial support, which may be lacking, the resulting loss of opportunity can lead to a loss of population and quality of life.

Fisheries, tourism, agriculture and (to some extent) mining have demonstrated important economic linkages in South Greenland, and these have been strongest when local businesses and labour markets are given priority in the consideration of inputs and service provisions.

Decisions that provide current economic relief to a local population can make it harder to achieve long-term sustainability. If immediate economic problems are solved with resource projects that depend on external support or have limited economic life, it may increase the long run challenge of achieving sustainability, and worsen the level of justice and inequality, and exacerbate existing power asymmetries. The challenge facing many Arctic communities is the trade-off between economic well-being and the environmental footprint. Hard decisions often need to be made between economic opportunities today or a more resilient and sustainable future in the long run. Unfortunately, existing power imbalances, the lack of leverage in negotiations, and the high degree of poverty in local communities mean that short-term planning tends to take precedence over long-term considerations, thereby worsening the resulting outcomes.

External ownership and control

The economic viability of northern communities is closely linked to what power the local level has when it becomes involved in processes of a potentially global scale. An important factor in achieving community viability is ownership rights to, or other forms of control over, natural resources (Aarsæther et al, 2004).

Large corporations can suppress local or regional entrepreneurship, drive out local competitors and inhibit the development of small-scale, local enterprises. These corporations often bring the necessary capital to finance, construct and operate large projects, and they help fill critical gaps in savings and investments, and entrepreneurship. This grants them substantial bargaining power that can minimize the net benefits for local communities. With limited bargaining power, local communities lose out on opportunities for value added, especially with economic leakages such as profits, resource rents and income; furthermore, the importation of intermediate products and services may create difficulties for the start-up and growth of private sector businesses locally (Huskey, 2010, 2011; Kruse 2010).

While the financial returns from resource extraction can be significant, so are the environmental and human costs. Continued large-scale exploitation activities are often met with strong opposition from local and environmental groups concerned about the damage to the environment and local livelihoods, as seen in the case of the rare-earth and uranium project once proposed in

Narsaq, South Greenland. Large-scale resource extraction can place a heavy burden on local infrastructure, services, housing and other facilities. It may draw on local labour otherwise engaged in traditional pursuits, who after the end of a project may find themselves permanently displaced. Traditional pursuits like hunting and fishing have a prominent role in the lives of the smaller communities in South Greenland, which also makes the pursuit of extractive industry complicated and risky.

Large-scale natural resource development, and especially fossil-fuel exploration and iron-ore and uranium mining, has been of specific interest. Many Greenlanders have seen it as a possible way forward toward greater independence and a more self-reliant economy (Poppel, 2018; Bjørst, 2017; Wilson, 2016; Nielsen, 2013; Hansen et al, 2016; Trump et al, 2018; Andersen, 2015; Rasmussen et al, 2014; Larsen and Huskey, 2020).

From the early 2000s and until more recently there was a policy move toward developing mining into a major industry, and considerable efforts were made to attract foreign mining companies (Nuttall, 2013). In 2013 the Greenlandic parliament repealed the country's zero-tolerance uranium policy and many began to see a proposed open-pit rare-earths and uranium mine in Kunnaersuit as critical for economic growth, self-sufficiency and greater independence (Bjørst, 2017, p 31). In 2021, however, the rare-earth and uranium mine was banned after intense opposition to the project and the results of published EIA and SIA reports in 2020.

While resource development in Greenland represents a potential key source of income and revenue, the process of exploiting non-renewable natural resources raises important questions on how to ensure that gains from these developments accrue to the residents of Greenland and that they do not have detrimental effects on human and environmental health. In fact, mining activities, oil exploration and large-scale industrial development plans have provoked considerable debates in Greenland about the significance and effects of such developments for society and the environment, the absence of appropriate public participation and consultation, decision making and regulatory processes, the shortcomings of social and environmental impact assessments, concerns for potential impacts on hunting, fishing and agriculture activities, and the influx of several thousand foreigners to work in the construction and operational phases of megaprojects (Nuttall, 2012, 2013; Larsen and Huskey, 2020).

In their case study focused on three proposed mines in South Greenland – the reopening of the Nalunaq goldmine near Nanortalik; the Kunnaersuit mine near Narsaq; and the Tanbreez rare-earth elements mine near Qaqortoq – Ramus Ole Rasmussen and Arild Gjertsen (2018) examined the local support for, and opposition to, mining and the increasing interest from foreign investors. As a way of considering these three mining proposals, the municipal council in Qaqortoq proposed a community strategy and action plan for mining activities to provide greater independence for Kujalleq from the central government, increasing local competencies and promoting a more diverse economy (Rasmussen and Gjertsen, 2018, p 132).

Resource extraction does not come without risks. In the context of Greenland and its desire by many for extractive industry projects, studies have highlighted the need to develop a broader based economy including non-resource extractive industries and activities, as well as the implementation of legislation and governance structures to handle the emerging resource economy, including clear principles, commitments and guidance on public consultations (Wilson, 2016, p 75).

The influence of external decision makers on the Arctic economy makes the pursuit of an equitable, just and sustainable economy more difficult. Many decisions important to the economies of the region and its communities are made outside the region and often reflect external conditions rather than local ones. Costs – such as environmental, socio-economic and cultural – ignored in the decisions by stakeholders outside the region, may interfere with and limit the ability of communities to support their local economy and daily living, and may subsequently become the source of inequities and power imbalances (Huskey and Southcott, 2016; Larsen and Huskey, 2010, 2020). Residents across the Arctic have responded to limited economic opportunities by moving. The state of education in Greenland presents significant labour market challenges and contributes to out-migration that eventually threatens the local economy.

Institutional change across the Arctic has increased the role of local residents in resource production decisions. The establishment first of Home Rule and later self-government in Greenland has given residents increased control of resource production and government spending.

A critical factor in creating a future that is just and equitable for all also includes how climate changes are being mitigated and what efforts are implemented to adapt to these changes to lower the economic fallout. Climate assessment reports by the Intergovernmental Panel on Climate Change (IPCC) have shown that climate change is escalating with cascading effects on economic sectors and human society. Moreover, climate impacts must be seen in the context of multiple stressors affecting the lives of local Arctic residents. There are benefits and costs (Hovelsrud et al, 2011; Larsen et al, 2014). While climate change may open the Arctic seas for transportation and continental shelf development, there is evidence that it may make it more costly to develop resources on land. Changes in flooding, permafrost and snow cover will increase the cost of production even in those areas with significant current resource activity. Warming may shorten the period during which ice roads allow for exploration and development activity on the tundra, while thawing permafrost may destabilize existing infrastructure, such as roads, pipelines and runways (Prowse et al, 2009). Climate change will challenge the Arctic's economic sustainability by increasing the cost of production of resources for the external market.

People living on the margin in remote settlements will be most affected by these climate changes and are also the ones with the least adaptive capacity, that is, with the least resources to be able to take the necessary action and who stand the most to lose from global change. Hence, climate change threatens to exacerbate existing inequalities.

Agriculture in South Greenland is an important growing sector that offers an alternative to extractive industry and a more internally resilient local community. This sector may also be able to benefit from a warming climate. In the period 2005 to 2015, when the prospects of mining dominated the Greenlandic media, Kujalleq residents were experiencing significant growth in fisheries, agriculture and tourism, raising the hopes for sustainable futures without extractive industries. Thus, the one mining proposal that faced fierce resistance in South Greenland was the proposed open-pit Kuannersuit rare-earth elements and uranium project near Narsaq. Not only was the project perceived as posing risks to Greenland's most significant horticulture and beef-cattle production area, it had prevented Narsaq from getting its name on the UNESCO's World Heritage List in 2017 as part of Kujataa (J.S.E. personal communication, 20 September 2021). Kujataa consists of five historical Norse and contemporary Inuit farming sites where the tourism industry had begun to grow before COVID 19, benefiting from the vicinity of the Narsarsuaq international airport.

As mentioned earlier, in 2021 Greenland's parliament passed legislation that bans uranium mining and has ceased the development of the Kuannersuit mine. Moreover, the current Greenlandic government has integrated several Sustainable Development Goals (SDGs) within its comprehensive strategy for agriculture. The strategy emphasizes the need for food security and shows that supporting and increasing the local production of domesticated plants and animals contributes to several of the SDGs, encourages the use of green energy, decreases the need for long-distance transportation of food products, and strengthens overall infrastructure in both rural and urban areas (Strategi for Landbrug 2021–2030, 2020).

In Kujalleq there are approximately 17,000 adult sheep on 37 sheep farms, and over 20,000 lambs are culled every year in the Neqi factory in Narsaq (Strategi for Landbrug 2021–2030, 2020). The number of sheep farms has been declining over recent decades, but this trend is not accompanied by a reduction in the number of sheep and is thus linked to a process of consolidation (Rasmussen, 2014) resulting from a subsidy scheme that was negotiated between the Sheep Farmers' Association and the government which encourages a minimum of 400 animals per farm as a way of increasing farmers' income (Landbrugskommissionens betænkning, 2014). Another cause of this trend has been difficulties in recruitment, as many young people did not want to raise livestock unless they were allowed to do so near the capital, Nuuk (M.S. personal communication, 20 May 2015). Livestock production in the Nuuk fjord was not made legal until 2017, and in 2018 a descendant of sheep farmers from South Greenland established a sheep farm near Kapisillit in Nuuk fjord, thus taking the first steps toward the Greenlandic government's ambitious aims of seeing 80 per cent of winter feed produced locally and 28,000 lambs slaughtered yearly (Strategi for Landbrug 2021–2030, 2020).

As the climate has become warmer, cattle production has emerged. Even though subsidy schemes do not yet extend to cattle production the number of beef cattle increased from 145 in 2015 to 292 in 2019 (Strategi for Landbrug 2021–2030, 2020). Horticulture is also a growing industry in South Greenland where in 2019 over 100 tons of potatoes were harvested at five family-owned farms. Located near Narsaq, the largest producer has asserted that with a bit of start-up support from the government, Greenlanders could become self-sufficient concerning various root crops and certain vegetables (Semionsen, 2019).

Climate change models in and of themselves point toward a positive future for Greenlandic agriculture, as predicted climate trends for the next few decades indicate that winters will be warmer and precipitation will increase during the summers. This could mean a longer growth period for root crops, vegetables and grasses, and less expenditure on watering systems – many farmers have had to irrigate their horticulture gardens as well as winter-fodder producing fields due to long dry spells during recent summers in South Greenland (Christensen et al, 2016).

Many experts contend that the prospects for Greenlandic agriculture will further improve if government support is directed at the most profitable farms and projects. It is argued that well-run farms do not get enough support to expand or improve their efficiency (Jevelund et al, 2016). Proponents of intensified agriculture want to see financial support that is more targeted, the strengthening of infrastructure, including roads and internet connections (Nymand, 2018), and a move toward diversification as well as the harnessing of green energy sources that would reduce costs and increase sustainability (Strategi for Landbrug 2021–2030, 2020).

Conclusion

Many parts of the Arctic region face challenges related to issues of justice, regional and local economic development, industrial production and large-scale resource extraction activities – some of which are structural and persistent. This includes remoteness and lack of accessibility, the high cost of production in the north, and human and other resource constraints; the consequences of environmental impacts from industrial development, and the negative spillover effects of industrial activity on local and Indigenous communities, culture and tradition; and the impacts of climate change. These and other socio-economic and environmental challenges exert their mark on the economic livelihoods of Arctic people and play important and growing roles in outcomes and decisions regarding resource allocation, resource use, ownership and control. These challenges can lead to both conflicts of interest and may exert harm on local environments and livelihoods, creating unjust and unequal opportunities for quality of life and standard of living. On the other hand, a renewed and increased emphasis on agriculture, tourism, fisheries and food processing might lead to more sustainable futures and greater independence for Greenland.

Study questions

1. What are some of the major challenges of economic development in the Arctic?
2. How does a lack of inclusion affect prospects for a more resilient future for Arctic residents?
3. What are the main sources of social and economic inequality in the Arctic?
4. What do you think is needed for a more equitable economic future?

Acknowledgements
This chapter has received funding from the European Union's Horizon 2020 research and innovation programme under grant agreement No 869327.

References
Aarsæther, N., L. Riabova, and J.O. Bærenholdt (2004) 'Community viability', in *Arctic Human Development Report (AHDR)*, Akureyri: Stefansson Arctic Institute, pp 139–54, [online], Available from: http://www.svs.is/en/projects/ahdr-and-asi-secretariat/ahdr-chapters [Accessed 30 December 2022].

Andersen, T.M. (2015) 'The Greenlandic economy – structure and prospects', *Economics Working Papers, 2014–15*. Department of Economics and Business Economics, Aarhus University.

Bjørst, L.R. (2017) 'Uranium – the road to "economic self-sustainability for Greenland"? Changing uranium-positions in Greenlandic politics', in G. Fondahl and G.N. Wilson (eds) *Northern Sustainabilities: Understanding and Addressing Change in the Circumpolar World*, Springer Polar Sciences, Cham: Springer, pp 25–34. https://doi.org/10.1007/978-3-319-46150-2_3.

Christensen, J.H., M. Olesen, F. Boberg, M. Stendel, and I. Koldtoft (2016) 'Fremtidige klimaforandringer i Grønland: Kujalleq Kommune', [online], Available from: https://www.dmi.dk/fileadmin/user_upload/Bruger_upload/Tema/Klima/15-04-01_kujalleq.pdf [Accessed 30 December 2022].

Danielsen, M., T. Andersen, T. Knudsen, and O. Nielsen (1998) *Mål og Strategier I den Grønlandske Erhvervsudvikling*, Nuuk, Greenland: Sulisa A/S.

Duhaime, G., A. Caron, S. Lévesque, A. Lemelin, I. Mäenpää, O. Nigai, and V. Robichaud (2017) 'Social and economic inequalities in the circumpolar Arctic' in S. Solveig Glomsrød, G. Duhaime and I. Aslaksen (eds) *The Economy of the North 2015*, [online], Available from: https://www.ssb.no/en/natur-og-miljo/artikler-og-publikasjoner/the-economy-of-the-north-2015 [Accessed 30 December 2022].

Hansen, A.M., F. Vanclay, P. Croal, and A.S.H. Skjervedal (2016) 'Managing the social impacts of the rapidly expanding extractive industries in Greenland', *The Extractive Industries and Society*, 3: 25–33.

Hovelsrud, G.K., B. Poppel, B. van Oort, and J.D. Reist (2011) 'Arctic societies, cultures, and peoples in a changing cryosphere', *Ambio*, 40: 100–10.

Huskey, L. (2010) 'Globalization and the economies of the North', in L. Heininen and C. Southcott (eds) *Globalization in the Circumpolar North*, Fairbanks: University of Alaska Press, pp 57–90.

Huskey, L. (2011) 'Resilience in remote economies: external challenges and internal economic structure', *The Journal of Contemporary Issues in Business and Government*, 17(1): 1–12.

Huskey, L., and C. Southcott (2016) 'That's where my money goes: resource production and financial flows in the Yukon economy', *The Polar Journal*, 6(1): 11–29.

Jervelund, C., N.C. Fredslund, and K. Jensen (2016) 'Målrettet støtte til det grønlandske landbrug', Departementet for Fiskeri, Fangst og Landbrug. Copenhagen Economics publication, [online], Available from: http://docplayer.dk/18195999-Maalrettet-stoette-til-det-groenlandske-landbrug.html [Accessed 30 December 2022].

Kruse, J. (2010). 'Sustainability from a local point of view: Alaska's North Slope and oil development', in G. Winther (ed.) *Political Economy of Northern Regional Development*, pp 55–72, [online], Available from: https://www.norden.org/en/publication/political-economy-northern-regional-development [Accessed 30 December 2022].

Landbrugskommissionens betænkning (februar 2014), [online], Available from: https://naalakkersuisut.gl/~/media/Nanoq/Files/Attached%20Files/Fiskeri_Fangst_Landb rug/DK/2016/Final_Rapport%20landbrug%202014_Chair_DK_pdf.pdf [Accessed 27 July 2022].

Larsen, J.N. (2002) *Economic Development in Greenland: A Time Series Analysis of Dependency, Growth and Instability*, Winnipeg: University of Manitoba Press.

Larsen, J.N. (2004) 'External dependency in Greenland: implications for growth and instability', Northern Veche, *Proceedings of the Second Northern Research Forum, Veliky Novgorod, Russia,* September, 19–22, 2002, edited by Jón Haukur Ingimundarson and Andrei Golovnov, Akureyri: Stefansson Arctic Institute.

Larsen, J.N. (2010) 'Climate change, natural resource dependency, and supply shocks: the case of Greenland', in G. Winther (ed.) *Political Economy of Northern Regional Development*, Vol. 1, Copenhagen: Nordic Council of Ministers, pp 188–98. doi: 10.6027/TN2010-521.

Larsen, J.N., and L. Huskey (2010) 'Material well-being in the Arctic', in J.N. Larsen, P. Schweitzer and G. Fondahl (eds) *Arctic Social Indicators*, Copenhagen: Nordic Council of Ministers, pp 47–66.

Larsen, J.N., and L. Huskey (2015) 'The Arctic economy in a global context', in B. Evengard, J.N. Larsen and Ø. Paasche (eds) *The New Arctic*, London: Springer.

Larsen, J.N., and L. Huskey (2020) 'Sustainable economies in the Arctic', in A. Petrov and J. Graybill (eds) *Arctic Sustainability: A Synthesis of Knowledge*. Abingdon: Routledge, pp 23–42.

Larsen, J.N., and A.N. Petrov (2020) 'The economy of the Arctic', in K.S. Coates and C. Holroyd (eds) *The Palgrave Handbook of Arctic Policy and Politics*, Cham: Springer, pp 79–95.

Larsen, J.N., O.A. Anisimov, A. Constable, A.B. Hollowed, N. Maynard, P. Prestrud, T.D. Prowse, and J.M.R. Stone (2014) 'Polar regions', in *Climate Change 2014: Impacts, Adaptation, and Vulnerability. Part B: Regional Aspects. Contribution of Working Group II to the Fifth Assessment Report of the Intergovernmental Panel on Climate Change.*, Cambridge, UK and New York: Cambridge University Press, pp 1567–612.

Nielsen, S.B. (2013) *Exploitation of Natural Resources and the Public Sector in Greenland: Background Paper for the Committee for Greenlandic Mineral Resources to the Benefit of Society*, Københavns Universitet. Baggrundspapirer / Udvalget for samfundsgavnlig udnyttelse af Grønlands naturressourcer, [online], Available from: https://research-api.cbs.dk/ws/portalfiles/portal/58811653/Soren_Bo_Nielsen_Exploitation_of_natural_resources_and_the_public_sector_in_Greenland.pdf [Accessed 30 December 2022].

Nuttall, M. (2012) 'Imagining and governing the Greenlandic resource frontier', *The Polar Journal*, 2(1): 113–24.

Nuttall, M. (2013) 'Zero-tolerance, uranium and Greenland's mining future', *The Polar Journal*, 3(2): 368–83.

Nymand, J. (2018) 'Agriculture, farming, and herding', in *Adaptation Actions for a Changing Arctic: Perspectives from the Baffin Bay/Davis Strait Region*. Arctic Monitoring and Assessment Programme (AMAP), Oslo, Norway, pp. 195–200, [online], Available from: https://www.amap.no/documents/doc/adaptation-actions-for-a-changing-arctic-perspectives-from-the-baffin-baydavis-strait-region/1630 [Accessed 30 December 2022].

Poole, G. (1990) 'Fisheries policy and economic development in Greenland in the 1980s', *Polar Record* 26(157): 109–18.

Poppel, B. (2018) 'Arctic oil and gas development: the case of Greenland', in L. Heininen and H. Exner-Pirot (eds) *Arctic Yearbook 2018: Arctic Development in Theory and in Practice*. Akureyri, Iceland: Arctic Portal, pp 1–32.

Prowse, T., C. Furgal, R. Chouinard, H. Melling, D. Milburn, and S.L. Smith (2009) 'Implications of climate change for economic development in northern Canada: energy, resource, and transportation sectors', *Ambio*, 38(5): 272–82.

Rasmussen, R.O. (2014) 'Multi-functionality as scenarios for land use development in the Arctic', in R. Weber and R.O. Rasmussen (eds) *Sustainable Regions – Sustainable Local Communities*. Nordregio Working Paper 2014: 2, Stockholm: Nordregio Publications.

Rasmussen, R.O., and A. Gjertsen (2018) 'Sacrifice zones for a sustainable state? Greenlandic mining politics in an era of transition', in B. Dale, I. Bay-Larsen, and B. Skorstad (eds) *The Will to Drill – Mining in Arctic Communities*, Cham: Springer, pp 127–49.

Rasmussen, R.O., G.K. Hovelsrud, and S. Gearheard (2014) 'Community viability and adaptation', in J.N Larsen and G. Fondahl (eds) *Arctic Human Development Report: Regional Processes and Global Linkages* (AHDR-II), Copenhagen: Nordic Council of Ministers, pp 427–78.

Semionsen, J. (2019) 'Vellykket høst: dobbelt så mange kartofler i år', *KNR – Greenlandic Broadcasting Corporation*, [online], Available from: https://knr.gl/da/nyheder/vellykket-h%C3%B8st-dobbelt-s%C3%A5-mange-kartofler-i-%C3%A5r [Accessed 22 January 2021].

Strategi for Landbrug 2021–2030 (2020) Naalakkersuisoq for Fiskeri, Fangst og Landbrug. Naalakkersuisut, [online], Available from: https://naalakkersuisut.gl/da/Naalakkersuisut/Departementer/Fiskeri-Fangst-og-Landbrug/Landbrug/Strategi. [Accessed 30 December 2022].

Trump, B.D., M. Kadenic, and I. Linkov (2018) 'A sustainable Arctic: making hard decisions', *Arctic, Antarctic, and Alpine Research*, 50: 1. doi:10.1080/15230430.2018.1438345.

Wilson, E. (2016) 'Negotiating uncertainty: corporate responsibility and Greenland's energy future', *Energy Research & Social Science*, 16: 69–77.

12

Seeing Like an Arctic City: The Lived Politics of Just Transition at Norway's Oil and Gas Frontier

Anna Badyina and Oleg Golubchikov

Introduction

The urban spaces of the Arctic stand at the crossroads of development trajectories that pose questions for a Just Transition. The economic development of the region has been characterized by its close links with industries. Many Arctic settlements have been developed based on a single industry which makes them sensitive to industrial shifts. The severe depopulation of many Arctic communities is a result of changes in economic structure and the degradation of many industries in the region. Many studies have induced a perception that the Arctic is a hostile environment to live socially and operate in economically (Hill and Gaddy, 2003). However, a number of Arctic cities and towns show markedly upward trajectories, not least driven by the oil and gas (O&G) activities and their high added value. The local proponents of this development argue that the Arctic can be an appealing environment for working and living despite its remoteness and harsh climate. For these communities, quality of life has been closely associated with collaboration with the O&G businesses for gaining local social benefits and creating public value.

The social significance of the O&G industry still rests upon multi-level coordination. In hydrocarbon-rich nations, the industry has been associated with a sizeable share of the national budget, which then finances the national welfare system (Norges Bank, nd). However, governments may also introduce specific requirements to ensure the O&G industry's contribution to the *local* community. The social significance at the level of local and regional development is still a complex, contested and evolving concept.

Many argue for a rethinking of the principles of the O&G participation in society, moving from abstract considerations (profit, welfare contribution and macroeconomic stability) to the actual socio-spatial practice of societal participation, including the development of urban and everyday life space. These principles are especially critical for the Arctic communities that have long struggled against their socio-spatial peripheralization.

The O&G sector pursues its social-environmental objectives in the form of corporate social responsibility (CSR), social licences to operate, or local public value creation. This responsibility may yet be interpreted in multiple ways. Largely, it depends on a company's profitability and is seen as a cost to a company – something which is financed as a *residual* activity. Companies are increasingly accepting the value of social dialogue for defining their social-environmental projects. However, the emphasis remains on specific outcomes rather than on defining the larger and holistic regional/community developments and their sustainability. This produces fragmented and time-limited impacts.

In the context of decarbonization and the green/sustainable transition, resource-rich Arctic communities face new dilemmas. On the one hand, they are compelled to be part of this transition; on the other, they will not have the capacity to develop and may degrade economically and socially if an alternative system to replace the O&G participation is not in place. This dilemma is further complicated by a lack of proper consideration for transition policies and the close relationship between the resource extraction industry and the development of an (urban) society. Transition risks creating *stranded assets* and, most importantly, *stranded communities*. This is stressed, for example, by the Just Transition Declaration adopted by fourteen governments and the European Commission at COP26 (International Trade Union Confederation, 2021; International Labour Organization, 2021).

Against this background, our chapter is set to explicate: (1) the close relationships between industrial and urban society development within the Arctic; (2) the possibility of integrating the urban society dimension in the *just* Arctic transition debate and practice; and (3) the significance of social dialogue that prioritizes inquiry into real-life community experiences and perceptions to improve practice and ethics. We will start with the latter point in order to outline our conceptual approach, and will then look at exploring these issues overall based on the case study of Norway's Hammerfest.

The everyday politics of social space

Arctic development is rarely considered as the nexus of *social relationship* embedded in the socio-spatial context. Yet, this approach can find support in seminal philosophical works on human practical existence and the ethics of this existence (Marx, 1845; Aristotle, 2009). Critical human geographers and urban sociologists argue that the strategies and processes of the development of human environment should be approached as the (social) production of space, as *socio-spatial practice*. Social space is not 'an independent material reality existing "in itself"; it is produced' (Lefebvre, 1991a). In other words, space should not be understood as a static category, but as a dynamic process rooted in a relational context (Goonewardena et al, 2008; Delbello and Nazha, 2018; Soja, 1980). The development of social space is embodied in actors and their thoughts, feelings, beliefs and visions; and embedded in physical and social contexts in which they operate and interact – their everyday life (Lefebvre, 1991b, 2002, 2005).

The human experience is central here. Social space is organized, but it is just a *moment of becoming* set in, and contingent on, actors' experience and perception of their (social) world in the practice of their everyday life. Each moment of becoming is a dynamic dialectical process of constructing social reality, which is not just about a concrete materiality, but 'a thought concept and a feeling – an "experience"' (Schmid, 2008, p 41). Materiality of human reality (physical settings or artefacts (legislation, institutions) do not exist 'without the thought that directs and represents [it], and without the lived experienced element, the feelings that are invested in this materiality' (Schmid, 2008, p 41).

This approach represents a 'materialist version of phenomenology' by emphasizing 'the process of social production of thought, action, and experience' rather than merely 'the subject that thinks, acts, and experiences' (Schmid, 2008, p 41). The emphasis is more on human becoming as a social, relational process – on *social praxis*. Praxis, understood as practical wisdom that is grounded in particular, perceptual and concrete experience (phronesis), is inseparably complementary to theory, understood as scientific knowledge that is generalizable, conceptual and abstract (episteme) (Igira and Gregory, 2009). Researchers emphasize 'the art of practice' and the epistemological importance of experience and reflection in defining the right course of action (Essays, UK, 2018). Humans act in accordance to how they make sense of the world around them and others in it. The knowledge they use is a specific 'moment' or 'event' of human experience (Bogusz, 2012, p 10). In this way, a successful practice is an *ethical* category in as much as it allows cultivating what Greek philosophers define as 'practical wisdom' – the skill to do the right thing, at the right time, for the right reason (McKay and McKay, 2020).

An individual with *practical wisdom* 'intuitively' grasps the specifics of the situation he/she acts in. It can only be generated through real-life experience. It depends on an individual's virtue (moral qualities), because what is good for someone depends on their individual qualities and life circumstances. According to this position, there is no rule for defining what is good in all cases because human life is not pre-given for humans but is constantly reshaped through human experience and action. What is *good* is always relative in time and space.

Having virtue and practical wisdom is thus a way of living in society; it requires interacting with others and negotiating with their perceptions and experiences (Lacewing, 2014). Practice offers humans a means to question and deliberate on the appropriateness of their standards of action and ways of thinking in real social situations. However, scholars have raised concern about practice being used not to better capture human (social) experience and perfect knowledge. Instead, practice is converted into 'non-practice' and 'the precondition for knowledge' by 'a singular practice of thinking, a practice of suspending practice' (Karsenti, 2007, p 140). As such, practice distances from the domain of experience where it originates and makes an effect.

Henri Lefebvre proposes a radically different approach to organizing social space, starting from the everyday, from lived experience. It is defined as *autogestion* – a process of democratic governance through which actors 'continually engage in self-criticism, debate, deliberation and struggle; it is not a fixed condition but a process of intense political engagement ... that must "continually be enacted"' (Lefebvre, 2001). It is 'a form of a grassroots political practice that "is born spontaneously out of the void in social life that is created by the state"' (Brenner, 2008, p 240). Autogestion is about 'qualitative transformation [of state power] into a non-productivistic, decentralised, and participatory institutional framework that not only permits social struggles and contradictions, but actively provokes them' (Brenner, 2008, p 240).

We attempt to operationalize these insights into the lived politics of practice/practical wisdom in understanding the relationships between industrial and urban society formations in the Arctic, with further reflections on (socially) Just Transitions. We do so through considering the historical experiences of one of the world's nethermost urban communities: Norway's Hammerfest. The analysis is based on an in-depth empirical study, which relied on secondary and primary data, a field trip to Hammerfest in October 2021, local focus groups and interviews, and further online interviews with multiple stakeholders representing local community, O&G businesses, and local and national government.

'We are building a society in Hammerfest'

Hammerfest is a small resource-based urban community located in Troms og Finnmark in the Norwegian Arctic. The recent development of the town offers fertile ground for research and policy. It contrasts with the oft-depressive representation of Arctic cities, demonstrating the possibility of developing an attractive thriving city with competitive qualities.

Formerly a declining centre of the fishing industry and a fishing port, since the early 2000s the town has shown an exceptional positive development which comes from the establishment of the O&G industry (the *Snøhvit* LNG plant and *Goliat* FPSO) (Benneworth, 2020). The skyline of the city has completely changed with new and refurbished community infrastructure, housing, roads and company offices. Now Hammerfest feels like an affluent and symbiotic city that offers interesting opportunities for work and everyday life.

Existing evaluations of the local and regional development have focused on a significant 'local value creation' or 'a local ripple effect', defined largely in terms of 'the development of the local supplier industry ... supplier contracts' and its specific content – 'a local office, opening for smaller vendors and alliances of vendors ... a local business incubator' (Holand et al, 2016). Research has also highlighted a set of necessary conditions to enable local value, such as management of expectations, requirements in contracts, close cooperation with the local authorities, and engagement with civil society (Holand et al, 2016).

These are all important accounts in detailing the effects and conditions of a remarkable economic change in Hammerfest. However, some fundamental 'procedural justice' aspects remain 'behind the scenes' – even if these are arguably at the root of what facilitates or hinders a quality transformation within the Arctic region. Specifically, the contingent nature of social practice in negotiating community benefits/public value is often blurred in the dominant accounts that tend to focus on *material* outcomes.

The Hammerfest development is a case of a 'local value creation' (or public value) based on collaborations between the O&G industry and Hammerfest community driven by local politics with a strong belief that 'people are the most important thing' and that the O&G developments within the Arctic cannot happen without 'building a society'. The companies and community have engaged in intense political debates, negotiations and 'fight' about what 'the local value creation' means for the city, people and their specific situations. The national government was proactive and specific at the key original stages of these processes. It has closely collaborated with the local and regional authorities, the trade union and industry. It is acknowledged that 'everyone just wanted to stand together and wanted it to happen ... it was like we all in this together, we are going to manage'. The national government 'has been on the same team, on the team for "building a society" not just economic gains' (the quotes in this and consequent paragraphs originate from our interviews with stakeholders and local experts in Hammerfest).

In this context, the O&G companies have chosen to use 'a non-traditional way' of operating their businesses by 'having a physical presence' in the community. They have established their operational offices in Hammerfest and required their technical suppliers to do the same. Having the O&G companies' offices in Hammerfest and all the associated infrastructure has been 'the most important thing' to happen through these negotiation processes. Economically, having the companies and their associated businesses in Hammerfest has provided local authorities with the required funds (property taxes) to refurbish the city. Socially, this has made the companies be closer to the community and thus to directly *experience* community life, develop necessary spatial sensibilities, and thereby become more accountable for this (social) experience. 'People are living and working here, so they are in not only in the heads, but in their hearts ... their children live here, their family ... they can feel and want to create ripple effects for themselves.'

It has been recognized by our respondents that the O&G companies have been keen on realizing meaningful changes within the community. They have engaged with local knowledge and experiences to understand the community needs and capabilities. They 'were spending lots of time travelling within Hammerfest and beyond to see what they can do for a ripple effect'. They thus took their 'social contract' seriously. Between 2002 and 2007, Statoil, one of the operating companies within Hammerfest, funded five non-company consultants who were working inside the Hammerfest municipality to make sure that the community develops proper plans. It was essential since the municipality at that stage had neither skills nor resources to develop the necessary capacities; it had only received about 5 million NOK from the government of Finnmark, which was not enough to build 'a society'.

By learning from the community experiences, the companies have developed a range of programmes to improve the life and prospects within the community. Vår Energy, for example, has focused on the following set of programmes:

- performing research and development activities;
- using local suppliers as much as possible;
- investing in projects and collaborations in primary, secondary and higher education to increase awareness and competency;
- supporting cultural projects to increase community attractiveness for existing and potential new residents; and
- conducting third party research that maps local ripple effects on a regular basis (Vår Energi, nd).

Establishing a local supply industry and all the infrastructure as 'a long-lasting industrial ripple effects and local value' has been another significant challenge in the industry–community negotiations. 'It has been a struggle for years to make the volumes big enough, to make sure they use suppliers from the North, because the companies have big contracts ... it has taken so many years, so much effort, so much money'. It can be argued that, although a local supply chain has now been established and is very 'thick' because of the O&G companies, it is still 'very, very fragile' if subjected to any politics that would make the companies leave or that would not recognize the necessity and difficulty of developing local suppliers within the northern circumstances.

This analysis suggests that conditions could have been different, and might still radically change in the future, with a different government strategy or company leadership in place, or even with a new macro-institutional requirement (the European Union's new Arctic policy, for instance). The outcomes and conditions are thus multi-scalar with multi-temporal contingent categories, which depend on the constellations of particular actors, their relationships between themselves and with the broader community and its material contexts – or social praxis (Smith, 1999). Simply put, much depends on who is at the decision-making table and on their specific experiences and perceptions.

Fundamentally, it is also about where decision makers and other actors are with their experiences and thoughts. This has to go back to people's education and how people are learning, what they want to do with their lives, how they see themselves versus the larger community, what they think their rights and responsibilities are, what their circumstances and capacities are, and what their social relationships are. It is not about building a cultural centre or developing a local supply chain, it is about who they are in their world/community. If you do not have a conversation around those issues, nothing is going to follow. All of this suggests an emergent, political, and 'lived' nature of engagement and negotiation process.

'The north needs to have a stronger urbanization'

Small northern cities are often discussed as peripheral areas besieged by multiple problems. This is largely associated with the global megatrends of urbanization (the concentration of people in larger urban areas) rather than with the structural dynamics of organizing urban life and space. There is a lack of national urban politics that recognizes the key role of cities in Arctic development.

Community experts and leaders in Hammerfest's transformations have argued that building an attractive city is even more important for 'building a society' than any other local value creation activities (providing jobs). The Hammerfest community has invested heavily in public infrastructure in a short period of time to provide for the needs and to support *optimism* within the community. It is acknowledged that people have started moving back to Hammerfest because 'it is now a city with growth, with the development, with a modernization that wasn't there before'.

People look for larger cities not only because of their agglomerative advantages but also for the quality of the living environment or the right combination of things for a good life (a choice of well-paid jobs, closeness to good schools, hospitals, cultural amenities and other social infrastructure). These qualities are not often present to the same extent in remote and 'harsh' Arctic areas. Local communities acknowledge that they have to fight against this global force, if they want to get 'enough bright minds, right hands and people living in the Arctic'. They say there is still 'a code to crack' which requires doing things differently. Many informants are passionate about the situation:

> 'We do have climatic issues, there is cold here, there is darkness in many parts of the months around the year and these are issues, problems, challenges that we can't facilitate … So our job in many ways is to show that there is possibility to live here, that there is a warmth here. That is a modern society in a cold climate. It is a continuous work that we have to do to show that … Many people when they hear about the Arctic, they think about glaciers and polar bears and frostbite, but that isn't actually the reality. We have power, we have electricity and there is light, sun and warm days. It is information that many people around the world don't actually know. There is culture, there is modern infrastructure and our job is to let people know – both in Norway and at the international scale – that there is a good livelihood to be made here.'

Local experts define further conditions that would make people move to the Arctic: the possibility to earn more and opportunities for travel supported by the top notch and cheap infrastructure.

'A stone-by-stone transition': building up from lived experience

The O&G industry/Hammerfest community collaborative processes towards 'building a society' and the positive outcome they have created are now being challenged by accelerated decarbonization and energy transition worldwide. Recently the new Arctic policy by the EU proposes to facilitate *green transition* in the Arctic, with intentions not to allow any further hydrocarbon reserve development in the region, nor to purchase such hydrocarbons if they were to be produced ('A stronger EU engagement for a greener, peaceful and prosperous Arctic', 2021).

This is undoubtedly an important advancement at the international level in addressing climate change. However, for the resource-dependent Arctic communities it means debilitating, if not devastating, consequences. As argued by the local community representatives, this bold decision will destroy the foundation for 'building a society' within the Arctic and actually prevent

Arctic communities from being a capable part of the green transition. This produces much grievance:

> '[The] EU has totally misunderstood how lives are here and how important the O&G sector has been for making sure people live here, so it is not like a museum. It's not like we pollute more here than in Europe. It is actually Europe, which is the biggest challenge. It is very interesting that they can just sit there and point to where we live. They just say stop everything and that means a consequence. We'll lose 25 per cent of our workforce ... We believe [that instead] if we build stone by stone, we also have the infrastructure, we have the competence, we have income, active politics and the supply industry. If we don't have that, we can't be part of this transition. We are doing it now. It has taken so much time, efforts and money. If we stop with O&G before we build up new industry, I think we will be like a museum.'

This demonstrates how the Arctic is essentially a *social space*. Any changes to this space would require connecting to others and their lived experiences. It would require recognizing that Arctic development is naturally contradictory and can only be appropriately organized through embracing and knowing emergent inconsistencies and disagreements. In the wider political discussions, the Arctic's green transition and sustainability is almost always narrowed down to the debates and initiatives against the detrimental environmental effects of the resource extraction sector. Few debates or critical assessments are offered regarding the resource extraction industry as a society-forming industry and of the patterns and effects of this social practice.

Arctic communities generally accept the importance of a low-carbon energy transition for their long-term sustainability and reveal that they are in a good position and have a unique opportunity to realize this shift based on the wealth of the cheap renewable resources they possess. Norway has been the largest producer of renewables in Europe. Its northern counties are among the largest counties in Europe producing renewables. Northern communities (including Hammerfest) have a much longer tradition of renewable resources than those of the O&G resources. Moreover, it is believed that a green transition opens new opportunities for an accelerated development of one of the key drivers of Arctic economy: local supplier industry. The latter can be an 'equal partner' or 'on the same starting line' with the O&G sector and its long-established supporting businesses.

The primary concern, however, is that if these economic opportunities are not developed in 'a better way' than has been done in relation to the O&G, this transition 'will not bring [the communities] any further, to ... developing good societies in the north'. It would make Arctic communities a 'harvesting place for companies' and development would happen elsewhere.

Resource-based communities like Hammerfest have been working on green activities in order to have 'other legs to stand on'. However, the main emphasis and effort remains on O&G operations. This is perhaps a natural inclination given the decision-making expertise and financial power of the O&G industry in contrast to the struggling local authorities in the north. This is largely about the specific 'social contract' – the socio-material relations and politics that communities have established with the O&G businesses. This 'contract' has revolved around building a modern urban society and has helped communities reverse the decline of their populations. Putting this system in place has taken 'so many years, so much effort, so much money'.

The O&G industry has been proactive in supporting the development of quality urban infrastructure within their hosting communities. They are also changing themselves: they are developing relevant 'green projects' within local communities. However, these developments can largely be defined as derivative of/dependent on the O&G operations. Furthermore, there is limited participation of the O&G business in or connecting with green or other economic activities within local communities that go beyond their core activities. This is because they see their business on its own, not as a relation. Their operation plan also recognizes a decommissioning plan as a physical act, detached from the context of their actions.

Based on our interviews, stakeholders directly or indirectly involved in O&G operations in the Arctic argue that decisions concerning O&G operations within the Arctic should consider an alternative system of 'building a good society', including financial mechanisms, key actors, competences, and their shared responsibilities. It is argued that there needs to be a strategy for communities in the north to develop *good societies*.

Green transition does not start with 'a blank paper' (in a vacuum), it develops within particular social geographies, structures, relations and circumstances that make Arctic communities. They are developed within the rather complex arrangements of human life. It is a continual process involving reflection, learning and improvement embedded within the system of life. It involves thinking of and managing 'different things at the same time'. The right approach to a green transition should be about building 'stone-by-stone' to allow local communities to create the necessary socio-material conditions for developing new and alterative industry competence and capacity.

Conclusion: The urban politics of a 'Just Transition' – three theses

A Just Transition is essentially practical
Arctic development is essentially about what Lefebvre recognizes as *the (social) production of space*. The key to a Just Transition lies in this process, in *the socio-spatial praxis* of Arctic initiatives, and knowing this *praxis*.

Central to Arctic development are 'human beings in their corporeality and sensuousness, with their sensitivity and imagination, their thinking and their ideologies; human beings who enter into relationships with each other through their activity and practice' (Schmid, 2008, p 29). In other words, Arctic transformation 'is based on the relationship of the subject to his or her [social] world and is embodied in the corporeality of this subject' (Schmid, 2008). Space is shaped through social relations, conflicts and negotiation of conflicts.

The existing literature has largely offered pessimistic stories (often focusing on environmental degradation). However, this literature does not explicate on what can be altered and how to make good choices. Such questions are particularly pertinent today when Arctic development is marked by contradictions, differences, inconsistencies and conflicts. This chapter has highlighted that one needs to return to classical philosophical thought on the *practical, experiential and experimental* nature of the human world and on the significance of *ethics* and *politics* in shaping it.

The critical evaluation of the Hammerfest development has demonstrated that the ethics and politics of the O&G, green or any other initiatives within the Arctic are key to ensuring they develop a long-term, just and sustainable Arctic. The ethics and politics of Hammerfest's progressive praxis have been essentially nurtured by non-traditional, proactive practices based on the collaboration of industry, state and local community. The Hammerfest case study has revealed that this practice has been principally grounded at the level of Hammerfest city and involved: (1) making O&G actors *experience* the reality of life in Hammerfest and be accountable for this *experience*, and (2) being open and willing to enhance the practice through others' perspective and experiences.

The epistemological significance of experience

The processes of 'local value creation' within Hammerfest demonstrate how transforming the Arctic region and communities and developing knowledge of what the good options are can find support in defining and responding to the region and real life community experience. This seemingly 'residual' knowledge may help Arctic communities develop from within and activate what can be defined as the 'self-production of space'. It is important that those *spaces* are recognized as 'catalysts for their own spatial logic' or of their own ideas of their reality, qualified to define and manage their own trajectories independent of the state and capital. This is not to say that there is no role for the state and capital; on the contrary, they need to assume a qualitatively different form of decision making. This should involve 'a non-productivistic, decentralized and participatory institutional structure' through which actors 'continually engage in self-criticism, debate, deliberation, and struggle' over their experiences and ways of organizing a social space (Brenner, 2008, p 240).

While Hammerfest has been significantly transformed through collaborations among the state, companies and the local community, this has been limited. Multiple internal and external factors have influenced development, but they have not been captured in their entirety in the formal negotiation processes. Although Hammerfest looks a modern town, those *unknown knowns* still facilitate uncertainty for the future. This chapter has discussed some of what may look like a rather messy mosaic of factors, including: the state's approach to the O&G sector's societal participation, the proactive political practice of local value creation, the O&G companies not often pushing for developments not in their direct interest, the role of major infrastructure development, national education system, property tax system, the problem of connectivity, and the problem of uneven regional and urban development.

We argue for a new epistemology of Arctic development – a new tradition advancing major socio-environmental transformations. This theory of knowledge builds from the argument that Arctic development represents a set of complex, conflictual and messy processes grounded in the reality of communities and shaped through multiple factors (external/internal, visible/invisible, structural/informal, global/local). Arctic development is not about a discrete project lasting for a certain period of time; it is not a fixed practice. It is largely about managing very dynamic processes by trying to identify and address multiple factors. It should be organized as a continuous process of negotiation, action, reflection and learning. It requires connecting among individuals, groups, organizations and institutions in their specific circumstances. Any initiative that aims to define the scope and magnitude of Arctic development once and for all deprives the Arctic region and its communities of the ability to adapt and improve. Some developments like Hammerfest can be used as an 'entry point' to demonstrate the importance and fundamentals of this epistemology.

The urban dimension of social praxis

Our chapter also reveals how the urban problematic is a critical component of the (social) production of space, of the socio-spatial dynamics that makes the Arctic region and communities. We show that building 'a good society' within the Arctic through industry–community collaborations has been connected to developing *urban qualities* or social infrastructure that accommodates the current and future population's needs and standards. These urban conditions are also critical for enabling a paradigmatic shift, particularly with respect to the *green* transition.

While the Arctic is a precious environment that needs to be preserved and cared for, where it represents a human habitat it is a highly urbanized space. Although this urbanized space is characterized by uneven development with many economically declining mining areas, it also has places with good, modern infrastructure.

We highlight the importance of integrating the *urban* as part of the Arctic development and the effects of different modes of organizing urban space. We can argue that urban development should start with the *urban* as *a way of life* and as *a web of life* and recognize it as 'social space' with 'social relations', and as a relational space structured across different dimensions and levels of human life.

This chapter demonstrates what progress has been made in Hammerfest with developing an urban society and attracting people living and working. However, there still remain many factors and conditions unconsidered, particularly those beyond the control of local actors. Any significant changes should integrate an active urban politics that aims to develop top notch urban qualities. Thus, politics should build on learning from positive urban practices within the region and embrace and work with structural dynamics that may hinder or otherwise facilitate positive development.

Study questions

1. Why is it important for key economic actors operating in a local community to understand the everyday local contexts not directly associated with their business?
2. In which way might this understanding be seen as a dimension of justice?
3. What are the practical lessons from the creation of public value out of the O&G industry in Hammerfest? What factors contributed to this value creation and capture?
4. How can the Arctic urban economies that rely on the O&G industry meaningfully engage with low-carbon energy transitions? What dilemmas do these communities and their host nations face?

References

'A stronger EU engagement for a greener, peaceful and prosperous Arctic' (2021) The European Union. Last modified 8 April 2022, [online], Available from: https://ec.europa.eu/commission/presscorner/detail/en/ip_21_5214 [Accessed 1 January 2022].

Aristotle (2009) *The Nicomachean Ethics*, translated and edited by W.D. Ross, with an introduction by R.W. Brown, Oxford: Oxford University Press.

Benneworth, P. (2020) *Universities and Regional Economic Development: Engaging With the Periphery*. Abingdon: Routledge.

Bogusz, T. (2012) 'Experiencing practical knowledge: emerging convergences of pragmatism and sociological practice theory', *Pragmatism and the Social Sciences: A Century of Influences and Interactions*, 2: 1–23. doi: 10.4000/ejpap.765.

Brenner, N. (2008) 'Henri Lefebvre's critique of state productivism', in K. Goonewardena, S. Kipfer, R. Milgrom and C. Schmid (eds) *Space, Difference, Everyday Life: Reading Henri Lefebvre*, New York: Routledge.

Delbello, L., and N. Nazha (2018) 'Edward Soja's socio-spatial dialectic and the simultaneous unfolding of time, space, and being', *Revista Científica OMNES*, 1(3): 185–201.

Essays, UK (2018) 'Pragmatism – Dewey and experiential learning lecture', Last modified 20 April 2022, [online], Available from: https://www.ukessays.com/lectures/education/approaches/pragmatism/ [Accessed 20 April 2022].

Goonewardena K, S. Kipfer, R. Milgrom, and C. Schmid (2008) *Space, Difference, Everyday Life: Reading Henri Lefebvre*. New York: Routledge.

Hill, F., and Gaddy, C. (2003) *The Siberian Curse: How Communist Planners Left Russia Out in the Cold*, Washington, DC: Brookings Institution Press.

Holand, K., M. Darell, and R. Rønning (2016) 'Hammerfest LNG Plant in Norway – significant local value creation', Presented at the *SPE International Conference and Exhibition on Health, Safety, Security, Environment, and Social Responsibility, Stavanger, Norway*. https://doi.org/10.2118/179285-MS.

Igira, F., and J. Gregory (2009) 'Cultural historical activity theory', in Y. Dwivedi, B. Lal, M.D. Williams, S.L. Schneberger and M. Wade (eds) *Handbook of Research on Contemporary Theoretical Models in Information Systems*, Hershey, PA: IGI Global, pp 434–54.

International Labour Organization (2021) 'ILO welcomes COP26 Just Transition Declaration', [online], Available from: https://www.ilo.org/global/about-the-ilo/newsroom/news/WCMS_826717/lang-en/index.htm [Accessed 1 January 2022].

International Trade Union Confederation (2021) 'Government at COP26 pledge support for just transition'. Last modified 7 May 2022, [online], Available from: https://www.ituc-csi.org/governments-at-cop26-pledge [Accessed 1 January 2022].

Karsenti, B. (2007) 'Une alternative au-delà du pragmatisme. La pratique en suspens', in M. de Fornel and C. Lemieux (eds) *Naturalisme versus constructivisme*, Paris: Éditions de l'EHESS, pp 133–40; cited in T. Bogusz (2012) 'Experiencing practical knowledge: emerging convergences of pragmatism and sociological practice theory', *Pragmatism and the Social Sciences: A Century of Influences and Interactions*, 2: 1–23. doi: 10.4000/ejpap.765.

Lacewing, M. (2014) 'Aristotle – practical wisdom', [online], Available from: http://cw.routledge.com/textbooks/alevelphilosophy/data/A2/Moral/PracticalWisdom.pdf [Accessed 1 January 2022].

Lefebvre, H. (1991a) *The Production of Space*, translated by Donald Nicholson-Smith, Oxford: Blackwell Publishing.

Lefebvre, H. (1991b) *Critique of Everyday Life. Volume I,* London: Verso.

Lefebvre, H. (2001) 'Comments on a new state form', translated by N. Brenner, *Antipode,* 33(5): 769–82.

Lefebvre, H. (2002) *Critique of Everyday Life. Volume II*, London: Verso.

Lefebvre, H. (2005) *Critique of Everyday Life. Volume III*, London: Verso.

Marx, K. (1845) *Theses on Feuerbach*, [online], Available from: http://www.marxists.org/archive/marx/works/1845/theses/index.htm [Accessed 1 January 2022].

McKay, B., and K. Mckay (2020) 'Practical wisdom: the master virtue', *Art of Manliness*. Last modified 3 June 2021, [online], Available from: https://www.artofmanliness.com/character/behavior/practical-wisdom/ [Accessed 1 January 2022].

Norges Bank (nd) 'About the fund'. Last modified 27 February 2019, [online], Available from: https://www.nbim.no/en/the-fund/about-the-fund [Accessed 1 January 2022].

Schmid, C. (2008) 'Henri Lefebvre's theory of the production of space: towards a three-dimensional dialectic', in K. Goonewardena, S. Kipfer, R. Milgrom and C. Schmid (eds) *Space, Difference, Everyday Life: Reading Henri Lefebvre*, New York: Routledge.

Smith, M.K. (1999) 'What is praxis?' in *The Encyclopaedia of Informal Education*, [online], Available from: http://infed.org/mobi/what-is-praxis/ [Accessed 28 November 2016].

Soja, E. (1980) 'The socio-spatial dialectic', *Annals of the Association of American Geographers*, 70(2): 27.

Vår Energi (nd) 'Ripple effects'. Last modified 7 May 2022, [online], Available from: https://varenergi.no/en/sustainability/social/local-value-creation/ [Accessed 1 January 2022].

Conclusion: Making Connections between Justice and Studies of the Arctic

Johanna Ohlsson and Corine Wood-Donnelly

Justice is for many a reiterative and ongoing process. To see where, for whom and how justice can be achieved begins by identifying both existing and potential future injustices that form the epicentre from where transformation can emerge. The work of this volume has intended to introduce justice to the conversation on development in and research on the Arctic, but also to flag injustice and to bring forth new ideas. In this conclusion, we discuss some of the key findings of the chapters, how the chapters relate and speak to each other, and the chapter culminates with a few ideas for further research. Here we are returning to the notions of justice and injustice, and we address how these concepts have been useful in the analyses in the preceding chapters.

This conclusion provides us with the opportunity to discuss how the themes, topics under study and the different aspects of justice coalesce in the volume. As has been made clear, the chapters make use of different types and understandings of justice as analytical tools as well as descriptors for various situations. For instance, this work addresses and problematizes both legal and social justice factors in several of these chapters. Most chapters focus on the issues of procedural and structural aspects of injustice in the Arctic and many discuss aspects of representation and recognition – or lack thereof.

We find both separate and overlapping understandings of justice throughout the volume, and several of the chapters speak towards one another, sometimes from a different aspect of justice thinking, sometimes with a different form, and even others from a different realm of justice.[1] What many of the chapters have in common is the acknowledgement of

---

[1] For further discussions of forms and realms of justice, see Ohlsson, J. and Przybylinski, S. (forthcoming 2023) *Theorising Justice: A primer for Social Scientists*, Bristol: Bristol University Press.

the critical and necessary potential of assessing the issues of development conditions in the Arctic through the lens of justice. This furthers a clear assumption based on the premise that one cannot solve one injustice by creating another. An important assumption that much of the reasoning builds upon is that when we present features of justice, issues of injustice also inevitably become apparent. Focusing on injustice or why something is unjust (rather than explicitly focusing on what is just) reveals important information and a more nuanced understanding of the circumstances. This, in turn, contributes to a more nuanced understanding of what justice is both in general and particularly within the context of the Arctic.

The chapters in this volume are situated within the broader context of justice literature in several ways, including traditional schools of justice, in their focus on specific features of justice and their explicit consideration of forms of justice. Given the emphasis on the liberal tradition of justice (Chapters 4, 8, 11 and 12), in substance there is more focus on freedoms and individual rights. Situated within the cosmopolitan tradition Chapters 3, 9 and 10 have an emphasis on the transnational and international aspects of rights and responsibilities. The contributions within critical approaches to justice (Chapters 1, 2, 3, 6 and 7) query relations of power and recognition, while the contribution using the capabilities approach (Chapter 5) emphasizes the relationship between empowerment and well-being.

The contributions of this volume also connect to existing features of justice found in the broader literature. For example, in a focus on intergenerational justice (Chapter 10), there is an ongoing discussion on the impacts of contemporary decision making on future generations. In the focus on Indigenous issues (Chapters 2, 8, 9, 10 and 11) there is attention drawn to issues of misrecognition and hierarchical inequality. Several chapters (6, 7, 10 and 11) position the environment as central to their concern of injustice and the position of the environment within decision-making processes.

Just Transition (Chapters 4, 9 and 12) takes into account a more systemic evaluation of trade-offs in responsibility and distribution of the effects of the green transition, while in the focus on energy justice (Chapter 5) there is concern for the distribution, production and consumption of energy resources.

A number of chapters within this volume connect to existing forms of justice, which address the modalities in which justice unfolds. For example, there are concerns about procedural forms of justice (Chapters 1, 3, 4, 5, 8 and 10) which evaluate processes and, in particular, inclusion within decision making and the formation of legal and political processes. The focus on recognitional forms of justice (Chapters 1, 2, 8, 9, 10, 11 and 12) considers who is included within processes or has a voice in decision making, or who suffers the consequences of decisions. Distributional forms of justice (Chapters 2, 6, 7, 8 and 11) are concerned with how resources, power or hazards are allocated across different hierarchies and groups within society.

There are at least three significant features of the Arctic that highlight its potency for justice scholarship: (1) the evidence of feedback loops emerging in climate change; (2) the perspective that the region is a resource base for economic development; and (3) the place that Arctic communities – especially Indigenous communities – have within this landscape. These features reveal that it is a time for reckoning in the distribution of harms and benefits, the decision-making procedures, and recognition of the role, rights and stakes that citizens and inhabitants have in both the past and the future of the Arctic. The chapters within this volume have introduced us to a variety of issues of injustices and provided us with perspectives from justice theorizing that may inform more just approaches to the region – drawing attention to aspects of the who, what, why, where and how of justice and injustice, first developed by Allison Jaggar (2009). Asking these five questions helps organize and describe an issue or context where justice or injustices are present, and to conduct analyses grounded in theories of justice.

A few themes cut across several of the chapters – one of the overarching themes is how issues of justice and injustice in the Arctic could, and sometimes should, be understood. For instance, Chapters 1 and 3 address the normative principles or standards helpful for assessing justice and taking responsibility for the effects of global climate change. While Chapter 1 centres on the organizational structure of the Arctic Council, Chapter 3 contributes a discussion on the centrality of a relational model of responsibility. Both ground their arguments in Critical Theory and draw on Frankfurt School accounts. Chapters 2 and 4 discuss other models for taking and distributing responsibility for injustices in the Arctic. Here, structural injustices and Iris Marion Young's (2011) five faces of oppression and domination, as well as the JUST framework and relational model for a better understanding of corporate social responsibility (CSR), are explored in the context of the Arctic. This contributes to a critical theoretical discussion towards our understanding of the Arctic. Chapter 5 furthers the discussion on the capabilities approach by expanding the theory to include collective capabilities. This also speaks to a relational approach, not far off from the one proposed in Chapter 3, yet the approach is from a different perspective and focuses on a different subject matter.

Chapter 6 critically examines the use of justice, with an explicit focus on environmental justice. The use of justice, according to this chapter, is often limited to the mainstreaming and signposting of justice approaches in environmental policies, instead of taking the unjust structures of capitalism and globalization seriously. It then frames its argument in the realm of responsibility. By risking a reduction of justice – and even more importantly – the injustices many people face, many policy documents adopt a version of environmental justice that in strategic ways abstracts from the actual injustices. This, instead, risks reproducing the very injustices that are supposed to be handled by not taking into account the historical and contemporary structures. Sharing a focus on environmental justice, Chapter 7 expands the previous US-dominated understanding of Sacrifice Zone, and tests its applicability to an Arctic context, explicitly in Norway.

Chapters 8, 9 and 10 centre on various topics, but all are related to Indigenous aspects of justice and injustice in the Arctic, primarily the Norwegian and Finnish Arctic. The features of justice they are highlighting are connected to representation, recognition and procedural justice.

Central points and avenues for future research

Many chapters offer important insights on practical processes, where issues of representation, recognition, responsibility and rights are pressing. This is often influenced by asymmetrical power relations. These asymmetries, we argue, need to be carefully assessed and taken into serious consideration when addressing development in the Arctic. As Darren McCauley (Chapter 4 of this volume) critically states, 'being responsible is not enough' when it comes to the roles and responsibilities of businesses in the Arctic. This is equally important for other sectors as well, not the least the public sector and public administration. Clearly, issues of justice and injustices are central parts of the development of sustainability agendas and the formation of just Arctic societies and territories. These issues must be treated as such.

What also becomes clear is that the just and ethical aspects of development in the Arctic are largely constituted by the relational and social aspects of power. This speaks to all chapters of the volume. It also shows that aspects of justice and injustice are important for future developments in the Arctic. This then strengthens our approach to initiating this field of study. We see this as the beginning of a very important conversation in the years to come.

The toolkit of justice theory provides richness, diversity and breadth in the options available to scholars. First, scholars can draw from the more traditional theories of justice, such as liberal, critical or cosmopolitan approaches, amongst many others. Second, there are established traditions that focus on a particular feature as the target of justice, such as climate, energy or space. Beyond this, scholars can choose from the forms of justice to investigate distributional, procedural or recognitional and retributional concerns at stake.[2] Depending on which justice tradition one employs, these different concerns will foreground in different ways. This makes studies of justice both complex and nuanced and at the same time provides opportunity for both narrow and broad investigations of injustice.

---

[2] For an exploration of various schools and fields of justice, see Ohlsson, J. and Przybylinski, S. (forthcoming 2023) *Theorising Justice: A primer for Social Scientists*, Bristol: Bristol University Press.

Our collective work in this volume has identified seven themes that appear promising for future research. The first theme is *recognition*. Several of the chapters discuss aspects of recognition, and this is (again) the start of a crucial discussion about Arctic development. Questions of who is seen, heard and listened to in debates, policy making, decision making, and planning of various initiatives are utterly important for addressing aspects of justice and injustice – both in scholarly work as well as in policy and business initiatives. Another separate but promising discussion is that of recognition (especially recognition of a variety of subjects – humans, non-human animals, ecosystems) in connection to issues of tolerance and respect. Other disciplines have had lively debates about these issues, but the discussion has, until now, overlooked the peoples, societies and ecosystems of the Arctic. With ecosystem services gaining attention in policy domains, studies are needed at the intersection of justice within ecosystems and the role of cultural ecosystems in prosperous and sustainable communities.

The importance of taking recognition seriously leads us to the second theme identified, *rights*. When stakeholders and rights holders are being appropriately included and listened to in various processes, more conflicts of interest and conflicts of rights may become increasingly apparent. Some of these already exist but have not always surfaced or have not been taken seriously enough. This will increase the need for (1) transparent negotiation processes, (2) enough time allocated for the consultation processes before the initiation of new projects (as well as critical assessments of the extension of old ones), and (3) knowledge of legal as well as local aspects in decision making and administrative processes at all levels. For instance, public officers at governmental agencies and municipalities must regularly review aspects of Indigenous rights to a larger degree than what currently seems to be the case. In addition, the political leadership at various administrative levels must create circumstances that properly allow for consultations with all affected people, even though industry requires a higher level of efficiency.

The third theme can also be considered to some extent as the starting point for any evaluation of justice and injustice. This speaks to the *vulnerability* present in various ways in the Arctic. The region includes vulnerable populations, vulnerable infrastructure and vulnerable ecosystems. This is made clear by several of the chapters in this volume. These chapters simultaneously serve as a venue for future research as more knowledge is needed. For instance, the connections between vulnerability and injustices in the Arctic seem to be important aspects for both theoretical and empirical exploration. This inevitably speaks to aspects of power and power asymmetries in various kinds of relationships and structures in the Arctic and of Arctic governance in particular.

A fourth theme that cuts across several of the chapters in this volume is *Indigeneity* and the challenges faced by Indigenous peoples, which is also an understudied topic. As the original peoples of the Arctic, it is the Indigenous communities that often have the most at stake in issues and questions of justice. There is much more to be understood towards retribution of historical injustice and restoration and recognition in postcolonial justice in order to co-create a better future. Discussions of Indigeneity are often embedded in issues of the exercise and protection of rights, participation in decision-making procedures and, more recently, surfacing in questions of Green colonialism. The question of subjectivity is increasingly contested in this domain and requires deliberate and deliberative attention.

A fifth theme is that most chapters in this volume are primarily people centred or *anthropocentric*. The majority of existing research relating to issues of justice and injustice – in the Arctic and generally speaking – is primarily centred on humans. However, issues of justice and injustice could also be related to other legal and moral subjects, resulting in novel accounts that, for instance, explore the rights of nature. The interconnected nature of the Arctic means that ecosystems and their non-human inhabitants (and their valuation) are related to the issues of justice. These discussions are striking with promising connections to several of the analyses in this volume, but there is also room for future work.

The sixth theme, *environmental justice*, has been woven from different geographical positions or perspectives with a focus on social aspects of justice within some traditions and with a focus on nature in others, and contributes to expanding discourses on the Arctic beyond the initial focus on the US. Some of the chapters in this volume make important contributions toward expanding the existing debates. For instance, Chapters 6 and 7 both contribute to the body of work on environmental justice. The concept of Sacrifice Zones contributes to the discussion of environmental justice in the Arctic, indicates that there is a clear gap in this area where more research is needed.

The seventh theme, on which others elsewhere have made important contributions, concerns the issue of *reconciliation* of Indigenous and minority groups in the Arctic. These are pressing issues in, but in no way limited to, communities in Canada, Sweden and Finland for instance. This doubtlessly has important consequences for issues of justice and injustices in the Arctic and reconnects to the first theme of recognition. The ongoing and recently initiated reconciliation processes sometimes relates to restorative and recognitional forms of justice, but the implications are yet to be seen.

What the work in this volume reveals is that the scope for exploring justice and the opportunities for removing injustices are many. What is required is the responsibility for this action to be assumed and for the work to begin in earnest. We hope that this volume acts as a prompt to this endeavour and becomes the catalyst for justice in the environment, societies and governance of the Arctic region.

References

Jaggar, A.M. (2009) 'The philosophical challenges of global gender justice', *Philosophical Topics*, 37(2): 1–15.

McCauley, D. (2023, in this volume) 'A JUST CSR Framework for the Arctic', in C. Wood-Donnelly and J. Ohlsson (eds), *Arctic Justice: Environment, Society and Governance*, Bristol: Bristol University Press.

Ohlsson, J. and Przybylinski, S. (forthcoming 2023) *Theorising Justice: A primer for Social Scientists*, Bristol: Bristol University Press.

Young, I.M. (2011) *Justice and the Politics of Difference*, Princeton, NJ: Princeton University Press.

Index

Page numbers in *italic* type refer to figures; those in **bold** type refer to tables.

A
Aarhus Convention 147
access, as an issue in CSR 57
Acker BP 52, **53**
AEPS (Arctic Environmental Protection Strategy) 26–7
affirmative principle (AP) 73, 74, **74**, 75, 76, 77
agriculture, South Greenland 158, 159, 162–3
AGW (anthropogenic global warming) *see* climate change
Agyeman, J. 83
Aleut International Association 14
'all affected' principle 10, 12, 13, 14, 16–17
Allard, C. 90
Alta struggle, Norway 111–12
Anchorage Declaration 142
AP (affirmative principle) 73, 74, **74**, 75, 76, 77
Appalachians, US 101
Arctic Athabaskan Council 14
Arctic Council 4, 24, 144, 146–7, 185
 governance and structural injustice 21, 24, 25–6, 26–7, 28, 30–1
 Monitoring and Assessment Programme 40–1
 and transnational theory of justice 8, 10–11, 14–15, 16, 17, 18
Arctic Environmental Protection Strategy (AEPS) 26–7
Arctic Environmental Responsibility Index 52, **53**, 61
Arctic exceptionalism 4, 8, 15–16, 17
Arctic Ice Project 141, 144
Arctic Region Declaration in Preparation for the Global Food Systems Summit (2021) 43, 46
Aristotle 69
Aurubis 117
autogestion 171

avoidable structural injustice 27
 see also structural injustice

B
Bay-Larsen, I. 105
Beowolf Mining PLC 85
Biedjovagge mine, Norway 109
Biersteker, T.J. 23
Biodiversity Act, Norway 113
Brännström, M. 90
Brenner, N. 171
Burch, K. 25

C
Cambou, D. 126
Canada 13
 Arctic Environmental Responsibility Index **53**
capabilities approach 5, 9, 68–72, 110, 119, 146, 184, 185
 and energy justice 72–3
 see also CCs (collective capabilities), and energy justice
capitalism 87–8
 production-reproduction nexus 90–1
carbon capture and storage (CCS) 140–1
carbon dioxide removal (CDR) 140–1
Carpenter, A. 57
CCS (carbon capture and storage) 140–1
CCs (collective capabilities) 67, 68–72, 185
 Arctic oil and gas development context 73–7, **74**
CDR (carbon dioxide removal) 140–1
CERD (Committee on the Elimination of Racial Discrimination), UN 86
Charles IX, King of Sweden 125, 131
Chickaloon peoples 38
'civic connections approach' model of responsibility 5, 36, 38, 46–7
Climate Action Tracker 45

INDEX

climate change 2, 3, 15, 17, 18
  impacts of 36–7, 38, 46–7, 74, 139, 161–2
  see also energy transition
climate justice 82
  geoengineering 145
  and responsibility 4–5, 36, 46–7, 185
    'civic connections approach' 5, 36, 38, 46–7
    co-responsibility 43–6
    ocean acidification 38, 40–2, 47
    relational model 36–8, 42, 45, 185
    wildfires 38, 39–40
Coggins, S. 10
collective capabilities *see* CCs (collective capabilities)
colonialism 10, 124, 125
  colonial legacies
    Forest Sámi peoples, Finland 6, 124–36
    Southern Greenland 6, 155–7
  green colonialism 92, 146, 187
  neo-colonialism 135
Comberti, C. 29
Committee on the Elimination of Racial Discrimination (CERD), UN 86
Committee on the Rights of the Child, UN 44
commodification 84, 87, 88, 91–2
commons, the, responsibility for 41–2
consent
  FPIC (free, prior and informed consent) procedure 43
  geoengineering and indigenous peoples 6, 140, 144, 145–6, 146–9
Convention on Biological Diversity, UN 42, 85, 147
COP26 169
corporate social responsibility *see* CSR (corporate social responsibility)
cosmopolitan justice 12, 16, 29, 51, 59, 184, 186
Council of Europe 90
Critical Raw Materials Resilience (European Commission, 2020) 85
Critical Theory 8, 9, 10, 185
CSR (corporate social responsibility) 5, 51–2, 53–4, 185
  Arctic Environmental Responsibility Index 52, **53**, 61
  energy companies in the Arctic 52, **53**, 54–7, 169
  JUST (Justice, Universal, Space, Time) framework 5, 52, 54, 57–9, 61, 185
  key principles of 60–1
cultural imperialism (Young's five faces of oppression) 21, 28, 29

D

Dahre, U.J. 89
Dale, B. 105
Day, M. 98, 99, 103, 104

Declaration on the Rights of Indigenous Peoples (UNDRIP), UN 43, 85, 143–4, 147
Del Casino, D.J. 25
deliberative structural injustice 27–8
  see also structural injustice
Denmark 13
  Arctic Environmental Responsibility Index **53**
  colonial legacies in Greenland 6, 155–7
  colonialisation of historical Sámi territory 124–5
Dewey, J. 70–1
dilution, as an issue in CSR 57
distributional justice 51, 59, 73, 82, 91, 110, 111, 184, 186
  Forest Sámi people, Finland 126, 132, 135
  geoengineering 144, 145, 146
  Sacrifice Zones 102–3
  Sámi people, Norway 118, 119, 120
Droubi, S. 58, 59
Dworkin, M. 72

E

ECHR (European Convention on Human Rights) 43, 45
ecosystem services 187
EEC (European Economic Community) 156
EIAs (Environmental Impact Assessments), Norway 113
embedding, as an issue in CSR 57
Endres, D. 98, 99, 100, 101, 102
energy justice 5, 82
  and capabilities approach 72–3
  CCs (collective capabilities) 67, 68–72, 185
  oil and gas industries in the Arctic 73–7, **74**
energy transition 66–7, 73–7
  see also green transition
Eni Norge 114, 115
Environmental Impact Assessments (EIAs), Norway 113
environmental justice 5, 81–3, 87–8, 184, 185
  Forest Sámi people, Finland 126
  Sweden 5, 81–2, 83–4, 91–2
    mineral mining 84–6
    production-reproduction nexus 90–1
    rights to the forest 88–90
    wind power 86–7
Equinor 52, **53**, 114
ERMA (European Raw Materials Alliance) 85
EU (European Union) 13
  Arctic policy 174, 175
  environmental policies 82
  Green Deal (2019) 5, 82, 92, 126, 136
  green energy transition 125–6
  mineral extraction policy 84–5
  Water Framework Directive 113

European Convention on Human Rights (ECHR) 43, 45
European Court of Human Rights 44n7, 45–6
European Economic Community (EEC) 156
European Landscape Convention (2000) 5
European Raw Materials Alliance (ERMA) 85
Evans, P. 71
Evenki peoples 38
exploitation (Young's five faces of oppression) 21, 28

F

Faden, R. 29
FAO (Food and Agriculture Organization), UN 38
Finland 13
  Arctic Environmental Responsibility Index **53**
  colonialisation of historical Sámi territory 124–5, 127
  Forest Sámi peoples 6, 124–7, 134–6
    current legal status 131–4, *133*
    difference from Mountain Sámi 128–30, *129*
    distributional justice 126, 132, 135
    environmental justice 126
    historical overview of rights 127–30
  Sokli mining project 132–3
Finnmark, Norway
  procedural justice in industry projects 109–10, 111–12, 120–1
    Goliat project 109, 111, 114–16, 119, 172
    Nussir mine 105, 109, 112, 116–18, 118–19, 120
fishermen's associations, Norway 114–15
five faces of oppression (Young) 4, 21, 28–30, 185
Food and Agriculture Organization (FAO), UN 38
Forest Revolt (Skogsupproret), Sweden 88
Forest Sámi peoples, Finland 6, 124–7, 134–6
  current legal status 131–4, *133*
  difference from Mountain Sámi 128–30, *129*
  distributional justice 126, 132, 135
  environmental justice 126
  historical overview of rights 127–30
Forst, R. 4, 8, 10, 11–12, 13, 15, 17–18
fossil fuels *see* oil and gas industries
Fox, J. 98, 99, 100, 103
FPIC (free, prior and informed consent) procedure 43
  geoengineering and indigenous peoples 6, 140, 144, 145–6, 146–9
Framework Convention on Climate Change, UN 42
Fraser, N. 83, 84, 90
Friends of the Earth 90

G

G50 report (Greenland) 155
Gállok/Kallal iron ore deposit, Jokkmokk, Sweden 85–6
Galtung, John 36
gas industry *see* oil and gas industries
Gazprom **53**, 55, 56
generality principle 10, 11
geoengineering 6, 139–40
  and indigenous peoples 140–4
    FPIC (free, prior and informed consent) procedure 6, 140, 144, 145–6, 146–9
  intergenerational justice 140, 144–6
GGR (greenhouse gas removal) 140
Gjertsen, A. 160–1
GLAN (Global Legal Action Network) 45
Global Food Systems Summit 43
Global Goals (2015), UN 82
Global Legal Action Network (GLAN) 45
global warming *see* climate change
Golgan peoples 38
Goliat project, Norway 109, 111, 112, 114–16, 119, 172
good life, the 69, 70, 71, 175
governance 2, 4
  structural injustice in 21–6, 30–1
    processes and consequences 26–7
    responsibility for injustice 27–30, 32
  and transnational theory of justice 12–15
green colonialism 92, 146, 187
green transition 135, 143, 145, 148, 169, 175–7, 179–80
  European Union 5, 92, 125–6, 136
greenhouse gas removal (GGR) 140
Greenland 154, 163–4
  balance of power in 157–9
  colonial period 155
  external ownership and control 159–63
  G50 report 155–6
  G60 report 156
  historical background 155–7
  Home Rule period 156–7, 161
  self-government 157, 161
Greenland Minerals and Energy 158
Grotuis, H. 41
Gwich'in Council International 14

H

Hamilton, T. 57
Hammerfest, Norway 6, 109, 112, 114
  Just Transition and the oil and gas industries 169, 171–7, 178–9, 180
Healy, N. 57
Hedges, C. 97, 98, 99–100, 103
Heffron, R.J. 57, 58
Heilinger, J. 22
Helgesen, V. 140
Holand, K. 172
Holifield, R. 98, 99, 103, 104

INDEX

human capabilities *see* capabilities approach
Human Rights Committee, UN 148
Hvinden, B. 105

I

ICCPR (International Covenant on Civil and Political Rights), UN 47, 127, 143, 147
ice geoengineering *see* geoengineering
Iceland 13
ICERD (International Convention on the Elimination of All Forms of Racial Discrimination), UN 127
ICESCR (International Covenant on Economic, Social and Cultural Rights), UN 143
Ilascu v Russia and Moldova (European Court of Human Rights) 44n7
ILO (International Labour Organization), Indigenous and Tribal Peoples Convention No. 169 (1989) 89, 127, 143, 147
Indigeneity 187
Indigenous and Tribal Peoples Convention No. 169 (ILO) (1989) 89, 127, 143, 147
Indigenous peoples 6, 13, 15, 184–5
  co-creation of research with 7
  marginalization of 28–9
  Permanent Participant status in the Arctic Council 14, 16, 21, 24, 26–7, 29, 30–1, 32, 146
  and Sacrifice Zones 100, 102, 103, 185
  self-determination, consent and participation rights 142–4
Indigenous Peoples Rome Declaration on the Arctic Regions Fisheries and Environment (2019) 38, 43
intergenerational justice 6, 184
  geoengineering 140, 144–6
Intergovernmental Panel on Climate Change (IPCC) 141–2, 161
International Covenant on Economic, Social and Cultural Rights (ICESCR), UN 143
International Convention on the Elimination of All Forms of Racial Discrimination (ICERD), UN 127
International Covenant on Civil and Political Rights (ICCPR), UN 47, 127, 143, 147
International Labour Organization *see* ILO (International Labour Organization)
international relations theory 4, 11, 13, 21
international system, structure of 23
Inuit Circumpolar Council 14, 26
Inuit peoples 38
IPCC (Intergovernmental Panel on Climate Change) 141–2, 161
Itelman peoples 38

J

Jaggar, A. 185
Jenkins, K. 72

JIMAB (Jokkmokk Iron Mines AB) 85
John III, King of Sweden 125
Jones, B.R. 72
Jonsson, H. 85, 86
JUST (Justice, Universal, Space, Time) CSR framework 5, 52, 54, 57–9, 61, 185
  key principles of 60–1
just sustainabilities 83
Just Transition 5, 6–7, 73n1, 82, 126, 146, 184
  oil and gas industries, Norway 6–7, 168–80
  see also energy justice; energy transition
Just Transition Declaration 169
justice 82–3, 183–4
  introduction and overview 1–8
  issues in 9–11
  liberal theories of 184, 186
  see also climate justice; cosmopolitan justice; distributional justice; energy justice; environmental justice; intergenerational justice; procedural justice; recognition justice; restorative justice; social justice; structural injustice; structural justice; transnational theory of justice
justification, right to in Forst's transnational theory of justice 8, 10, 11, 16

K

Karsenti, B. 171
Kebneskaise Glacier 141
Keutsche Group 141, 146
KGH (Kongelige Grønlandske Handel) 155
Kiruna, Sweden 141, 146
Kohn, M. 124
Kratochwil, F.V. 25
Kujalleq, Greenland 157, 162–3
Kvalsund, Norway 112
Kyoto Protocol, UN 42

L

LaBelle, M.C. 59
Lefebvre, H. 170, 171, 177
Lerner, Steve 98, 99, 103
liberal theories of justice 184, 186
'local value creation' 172, 173, 174, 178–9
Lukoil 55–6

M

Manteaw, B. 58
marginalization
  and Sacrifice Zones 100
  Young's five faces of oppression 21, 28–9
marine pollution, Norway 113, 114–15, 116–17
Markbygden 1101 wind farm, Sweden 86–7
Massingham, P. 69
McCauley, D. 57, 58
MCE (Ministry of Climate and Environment), Norway 117

McKeown, M. 26, 27
Metsähallitus, Finland 126, 133
Mineral Act, Norway 113–14
mining
  Forest-Lapland region, Finland 126, 132–3
  North Sea deep sea mining 47
  Norway 109–10, 112, 113–14
    Nussir mine 105, 109, 112, 116–18, 118–19, 120
    South Greenland 158, 159–61
  see also Sacrifice Zones
Mining Inspectorate, Sweden 85
Ministry of Climate and Environment (MCE), Norway 117
misrecognition 184
  Sámi people, Norway 118–19, 120
Mountain Sámi, Finland 128–30, *129*
mountain-top removal 96, 98, 103

N

Nanortalik, Greenland 157, 158, 160
Narsaq, Greenland 157, 158, 160–1, 162–3
Nayoga Protocol 147
NEA (Norwegian Environment Agency) 117
neo-colonialism 135
neoliberal imperialism 23
NGOs (non-governmental organizations) 113, 114, 141
Nickel Mountain AB 86
non-Arctic states 13
  Observer role in the Arctic Council 21, 24, 25, 32
Nordic political welfare model 105
North Atlantic, tensions in 9
North Sea, deep sea mining 47
Norway 13
  colonialisation of historical Sámi territory 127
  CSR (corporate social responsibility) practices 52, 53, **53**
  living conditions 105–6
  marine pollution 113, 114–15, 116–17
  mining 109–10, 112, 113–14
    Nussir mine 105, 109, 112, 116–18, 118–19, 120
  oil and gas industries 6–7, 109–10, 112
    Goliat project 109, 111, 112, 114–16, 119
    Just Transition and social issues 6–7, 168–80
    legal framework 112–13, 114
  procedural justice issues in Arctic industry projects 109–20
  renewable resources 109, 119, 120, 176
  Sámi peoples 6
    distributional justice 118, 119, 120
    misrecognition 118–19, 120
    'Norwegianization' policies 111–12
    procedural justice issues in industry projects 109–10, 111–12, 113–20

Norwegian Environment Agency (NEA) 117
nuclear waste 97, 98, 102
Nussbaum, M. 9, 68, 72
Nussir copper mine, Norway 105, 109, 112, 116–18, 118–19, 120

O

Observers, of the Arctic Council 24, 25, 29
ocean acidification 38, 40–2, 47
oil and gas industries 5, 9, 66–8, 71–2
  CCs (collective capabilities), and energy justice 73–7, **74**
  Norway 6–7, 109–10, 112
    Goliat project 109, 111, 112, 114–16, 119
    Just Transition and social issues 6–7, 168–80
    legal framework 112–13, 114
    procedural justice issues 109–20
  price volatility 67, 76
  public-private collaborations 67–8
  see also CSR (corporate social responsibility)
Onuf, N.G. 29
Osterhammel. J. 124
Ottawa Declaration 1996 14, 144
  see also Arctic Council
Overland, I 52, 53, **53**

P

Paris Agreement (2015), UN 47, 66, 76, 139
participation 91, 110–11
  see also FPIC (free, prior and informed consent) procedure; procedural justice
particularity *see* Arctic exceptionalism
PBA (Planning and Building Act), Norway 113
peatlands, and carbon storage 39
Permanent Participants, of the Arctic Council 14, 16, 21, 24, 26–7, 29, 30–1, 32, 146
Petro Arctic 115
Petroleum Act, Norway 113
Pettit, P. 37
Planning and Building Act (PBA), Norway 113
Poland 57
Pollution Act, Norway 113
positionality, in research 7
powerlessness (Young's five faces of oppression) 21, 28, 29
Powers, M. 29
PP (prohibitive principle) 73, 74, **74**, 75, 76, 77
practical wisdom 69, 170–1
practice 70–1
praxis 170, 177
presumptive responsibility principle 44–5
price volatility, oil and gas industries 67, 76
procedural justice 6, 14–15, 51, 59, 73, 82, 110, 111, 183, 186
  Forest Sámi people, Finland 126, 132, 135
  Hammerfest, Norway 172

INDEX

Sacrifice Zones 101–2
Sámi people and Norwegian Arctic industry projects 109–20
wind power in Sweden 87
prohibitive principle 73, 74, **74**, 75, 76, 77
property rights 41–2
proportionality principle 17, 47
public-private collaborations, oil and gas industries 67–8
pure structural injustice 27
see also structural injustice

Q
Qaqortoq, Greenland 157

R
rare-earth mining, South Greenland 158, 159–61, 162–3
Rasmussen, R.O. 160–1
Rawls, J. 21
realism, in international relations theory 11, 13
reciprocity principle 10, 11, 17
recognition justice 10, 14, 51, 59, 73, 82, 91, 110, 184, 186–7
Forest Sámi people, Finland 126, 132, 135
Sámi people, Norway 118–19
Reddy, K. 124
Reindeer Herding Act, Norway 115
reindeer husbandry 86, 88–9, 115–16, 117–18, 119, 120, 128, 129–30, 131–2
Reindeer Husbandry Act, Sweden 130
Reinert, H. 101, 105
renewable resources
Norway 109, 119, 120, 176
Sweden 86–8
responsibility 4–5, 36, 58
'civic connections approach' 5, 36, 38, 46–7
and climate justice 4–5, 36, 46–7, 185
co-responsibility 43–6
ocean acidification 38, 40–2, 47
relational model 36–8, 42, 45, 185
wildfires 38, 39–40
for the commons 41–2
presumptive responsibility principle 44–5
relational model of 36–8, 42, 45
see also CSR (corporate social responsibility)
restorative justice 51, 59, 73
right to justification, in Forst's transnational theory of justice 8, 10, 11, 16
rights, concept of 3, 4
rights, of Arctic peoples 6, 37–8, 60
Robeyns, I. 68
Rönnbäck/Rönnbäcken, Storuman, Sweden 86
Royal Geographical Society 140

Russia 13
colonialisation of historical Sámi territory 124–5, 127
invasion of Ukraine 10, 47, 66, 76
oil and gas industries 67
CSR (corporate social responsibility) practices 52, **53**, 54, 55–6
relationship with US in Arctic region 9–10
Russian Association of Indigenous Peoples of the North 14

S
Sacco, J. 97, 98, 99–100, 103
Sacrifice Zones 5, 96–8, 106, 185
activism 103–4
distribution of benefits and burdens 102–3
environmental impacts 98–9
Nordic Arctic justice studies 104–6
power and interests 100–2
socio-economic characteristics 99–100
SAI (stratospheric aerosol injection) 140, 141, 144, 145–6
see also geoengineering
Sakha peoples 38
Sámi Act, Finland 132
Sámi Council 14, 43–4
and geoengineering 141–2
Sámi Parliament 85, 112, 113, 115–16, 117
Sámi Parliament Act, Finland 130, 132–3
Sámi peoples 38
Forest Sámi, Finland 6, 124–7, 134–6
current legal status 131–4, *133*
difference from Mountain Sámi 128–30, *129*
distributional justice 126, 132, 135
environmental justice 126
historical overview of rights 127–30
and geoengineering 142–6
FPIC (free, prior and informed consent) procedure 6, 140, 144, 145–6, 146–9
Norway 6
distributional justice 118, 119, 120
misrecognition 118–19, 120
'Norwegianization' policies 111–12
procedural justice issues in industry projects 109–10, 111–12, 113–20
Sweden 85–6, 87, 128
rights to the forest 88–90
Sápmi 88, 111, 119, 125, 144, 145
Sardo, M.C. 53
Schmid, C. 170, 178
SCoPEx (Stratospheric Controlled Perturbation Experiment) 141
Scott, Rebecca R. 97, 98, 99, 101, 102
sea jurisdictions 13
Sen, A. 9, 37–8, 68, 72, 146
Settlement Bill of Lapland, 1673 125, 131

195

Siberia, wildfires 39
Sidortsov, R. 73
Simon, M. 26
Skorstad, B. 105
Snøhvit project, Norway 109, 114, 115
Social Constructionism 4, 21
social praxis 170
social space 169–71, 177–8
Sokli mining project, Finland 132–3
solar radiation management (SRM) 6, 140, 141, 142, 146
 see also geoengineering
South Greenland *see* Greenland
Sovacool, B. 72, 77
sovereignty 23, 25
 principle of 10, 13, 16
SRM (solar radiation management) 6, 140, 141, 142, 146
 see also geoengineering
stakeholders
 and CSR (corporate social responsibility) 54–5, 56–7
 Sámi people, Norway 117–18
states
 and the international system 23, 31
 role in the Arctic Council 21, 24
Statoil 173
Storheia windfarm, Norway 145
stranded assets 67, 169
stranded communities 169
stratospheric aerosol injection (SAI) 140, 141, 144, 145–6
 see also geoengineering
Stratospheric Controlled Perturbation Experiment (SCoPEx) 141
structural injustice in governance 4, 21–6, 30–1, 185
 processes and consequences 26–7
 responsibility for injustice 27–30, 32
structural justice 21–2, 183
subjectivity, in research 7
sub-Saharan Africa 73
Suopajärvi, L. 102
Sveaskog, Sweden 89, 90
Sweden 13
 Arctic Environmental Responsibility Index **53**
 colonialisation of historical Sámi territory 124–5, 127
 environmental justice 5, 81–2, 83–4, 87–8, 91–2
 mineral mining 84–6
 production-reproduction nexus 90–1
 rights to the forest 88–90
 wind power 86–7
 geoengineering 141
 landscape and tourism 90–1
 Sámi peoples 85–6, 87, 88–90, 128
Swedish Geological Survey 84

Swedish Mineral Law 84
Swedish Mineral Strategy 84

T

Tarakegne, B. 73
Thien, D. 25
Thorwaldsson, K.-P. 84
Tidholm, Po 92
Total 52, **53**
tourism
 Greenland 158, 159, 162
 Sweden 90–1
trade routes 13
Trainor, S.F. 97
translation, as an issue in CSR 57
transnational theory of justice 4, 8–12, 17–18
 Arctic exceptionalism 8, 15–16
 Arctic governance 12–15
 assessment of 16–17

U

Ukraine, Russian invasion of 10, 47, 66, 76
Ulchi peoples 38
UN (United Nations) 90
 CERD (Committee on the Elimination of Racial Discrimination) 86
 Committee on the Rights of the Child 44
 Convention on Biological Diversity 42, 85, 147
 FAO (Food and Agriculture Organization) 38
 Framework Convention on Climate Change 42
 Global Goals (2015) 82
 Human Rights Committee 148
 ICCPR (International Covenant on Civil and Political Rights) 47, 127, 143, 147
 ICERD (International Convention on the Elimination of All Forms of Racial Discrimination) 127
 ICESCR (International Covenant on Economic, Social and Cultural Rights) 143
 Kyoto Protocol 42
 Paris Agreement (2015) 47, 66, 76, 139
 SDGs (Sustainable Development Goals) (2015) 5, 162
 UNDRIP (Declaration on the Rights of Indigenous Peoples) 43, 85, 143–4, 147
 UNHRC (United Nations Human Rights Council) 143
 Universal Declaration of Human Rights 85
Universal Declaration of Human Rights, UN 85
uranium mining, South Greenland 158, 159–60, 162–3
urbanization 174
US (United States) 13, 57
 Arctic Environmental Responsibility Index **53**

relationship with Russia in Arctic region 9–10
Sacrifice Zones 97, 98, 100, 101, 102, 103

V

value creation, collective politics of 68
Vår Energi 113, 114, 173
violence (Young's five faces of oppression) 21, 28, 29–30

W

Wagner, M. 57
Weber, C. 23
Wendt, A. 23, 26, 30

West Virginia, US, coal mining 98, 102, 103
wildfires 38, 39–40
wind power 57
   Finland 134–5, 136
   Norway 109, 119, 120, 145
   Sweden 86–7

Y

Young, I.M. 4, 21, 22, 23, 24, 25, 26, 27, 28, 185
Young, O. 30
Ypi, L. 22
Yucca Mountains 100
Yukaghir peoples 38